STO

FRIENDS OF ACPL

684
Bla
49 easy-to-build plywood
projects

DO NOT REMOVE
CARDS FROM POCKET

ALLEN COUNTY PUBLIC LIBRARY

FORT WAYNE, INDIANA 46802

You may return this book to any agency, branch,
or bookmobile of the Allen County Public Library

DEMCO

49 Easy-to-Build Plywood Projects

Percy W. Blandford

TAB BOOKS Inc.
Blue Ridge Summit, PA

Allen County Public Library
Ft. Wayne, Indiana

FIRST EDITION
FIRST PRINTING

Copyright © 1990 by TAB BOOKS
Printed in the United States of America

Reproduction or publication of the content in any manner, without express permission of the publisher, is prohibited. No liability is assumed with respect to the use of the information herein.

Library of Congress Cataloging-in-Publication Data

Blandford, Percy W.
  49 easy-to-build plywood projects / by Percy W. Blandford.
      p.   cm.
  ISBN 0-8306-7344-X   ISBN 0-8306-3344-8 (pbk.)
  1. Plywood craft. I. Title. II. Title: Forty-nine easy-to-build plywood projects.
  TT191.B53 1990
684'.08—dc20   89-48462
                                                                           CIP

TAB BOOKS offers software for sale. For information and a catalog, please contact TAB Software Department, Blue Ridge Summit, PA 17294-0850.

Questions regarding the content of this book should be addressed to:

Reader Inquiry Branch
TAB BOOKS
Blue Ridge Summit, PA 17294-0214

Acquisitions Editor: Kimberly Tabor
Book Editor: Kathleen E. Beiswenger
Production: Katherine Brown

Contents

Introduction *vii*

1 Introducing Plywood *1*
    Sizes *3*
    Quality *4*
    Edges *6*
    Joints *7*

2 Tables and Stands *9*
    Take-down Coffee Table *11*
    Coffee Table *14*
    Pedestal *19*
    Flap Table *24*
    Folding Desk *27*
    Greenhouse Plant Stand *31*
    Bed Table *35*
    Game Table *37*
    Corner Table *41*
    Bedside Extending Table *45*

3 Seats and Stools *49*
    Foot Stool *51*
    Child's Chair *55*
    Armchair *59*
    Side Chair *61*
    Outdoor Chair *66*
    Bench Seat *70*
    Dresser Seat *72*

4 Containers *77*
  Chest *79*
  Umbrella Stand *85*
  Tool Tote Box *89*
  Rolling Tilt Bin *92*
  Tabletop Containers *95*
  Kitchen Tray *98*
  Hot Pads *100*
  Cabinet *102*
  Box/Roll Holder *107*

5 Racks and Shelves *111*
  Souvenir Rack *113*
  Magazine Rack *114*
  Take-down Bookshelves *117*
  Vegetable Rack *119*
  Hat and Coat Rack *123*
  Picture Frame *126*
  Floor Bookcase *128*
  Drafting or Macrame Desk *131*
  Modules *134*

6 Toys *137*
  Doll's Crib *139*
  Swan Rocker *140*
  Ring Game *143*
  Dollhouse *145*
  Doll Stroller *149*
  Toy Box/Seat *152*
  Wheelbarrow *157*

7 Outdoor Projects *161*
  Barbecue Table/Trolley *163*
  Stacking Seed Boxes *167*
  Plant Pot Stand *169*
  Yard Cart *172*
  Recliner *176*
  Yard Table *179*
  Garden Kneeler/Toolbox *182*

Index *188*

Introduction

In the modern woodworking world, plywood is very much a fact of life. It is used increasingly in furniture and other wooden assemblies and in places where solid wood might have been used. It is used to make structures that are impossible or difficult to make with solid wood. Its advantage is that it is available in large flat sheets of uniform thickness, with attractive surfaces when required. Trying to achieve similar effects to these panels in solid wood would involve much more work, and the result might split or warp in use.

It is easy to use plywood. Much of the work is quick, and anyone, whatever their woodworking skills, can expect to obtain good results. For many constructions, plywood is usually cheaper than solid wood.

In addition to using plywood with solid wood, you can make many projects almost entirely of plywood, and that is what this book is about. There are many things to make, all of which can stand comparison with those made in other ways, so you can be proud of your work. Modern plywood is structurally sound and the glues used are damp- or water-resistant, so your plywood project will have a long life and be able to stand up to normal or even rough use.

Working with plywood as the main construction material is little different from working with solid wood. In many ways it is easier. It is possible to make many of the projects in this book with just a few basic hand tools. Because there are no large surfaces to plane, you only need a plane for edges. Panels are thin enough to be sawn by hand. If you can also drill holes and drive nails and screws, you can start making worthwhile projects. If, however, your woodworking equipment is more extensive, you will find uses for it. Power tools, such as a table saw, will take less effort and make accuracy easier to obtain. Because squareness is important in many assemblies, you will need marking out and testing tools, but you can always start with the squareness of the corner of a standard sheet.

The 49 projects in this book range from small to large, simple to more advanced, and inexpensive to more costly. There is something for everyone. You can produce simple things at little cost by using scraps, or you can tackle more involved projects by using several sheets. Most designs are intended for the amateur woodworker or the one-off professional craftsman, but many items can be produced in short or long runs for sale. Because plywood lends itself to quantity production, sheets can be cut economically and waste kept to a minimum.

This is not a book on woodworking techniques, and some basic skill is assumed. When you select a project to make, refer to chapter 1 for some tips on selecting and dealing with plywood.

May you get considerable satisfaction from working with plywood and making some of the projects in this book.

Note

All sizes are in inches unless marked otherwise. In the Materials Lists, some dimensions are oversized to allow for trimming.

1

Introducing Plywood

Plywood is a modern manufactured board with a great range of possibilities and only a few limitations. Its greatest advantages over solid wood are the availability of broad panels, uniform strength, and comparative lightness. Early plywoods were made with glues that were unreliable, and panels sometimes delaminated. Since World War II, glues have developed to the point where there should never be any fear of layers separating.

Plywood is made of three or more layers of thin wood laid alternately square to each other and glued. The plies or veneers in most plywoods are cut by a broad knife while the log rotates on a machine like a lathe. This means that any grain markings are not like those of a board cut in the usual way across a log. In most cases, the surface of a piece of plywood does not have a very pronounced grain appearance because the markings due to spring and fall wood are not cut across, but around.

If you match plywood and solid wood of the same species, you might get similar coloring, but do not expect the same grain appearance. It is possible to buy plywood with surface veneers applied, which are cut across the grain to give the appearance of solid wood. On a wide panel, however, there might be joints in several places, as there is a limit to the widths that such veneers can be cut across a log. Because of the circumferential cutting of the veneers of ordinary plywood, the veneers can be of almost unlimited cross-grain lengths.

Sizes

Most sheets of plywood are 48 inches × 96 inches. There are other sizes, but this size is what is commonly available. It is as big a piece as a man can expect to span with his arms and carry. Your supplier will cut you smaller pieces, but you will pay proportionately more. If you think you can eventually use the extra plywood, buy the sheet, even if it is more than you need for the current job. Your supplier might be willing to cut it through to more convenient sizes, providing you buy it all.

Many thicknesses are possible, but you might find that those available locally are limited. Softwood plywood cannot be made very thin, but hardwood plywood can be any thickness, right down to "aircraft" plywood of three veneers in a total thickness of 1 mm or $1/25$ inch.

Most plywood is available in $1/8$-inch steps of thickness. For most constructional work, such as the projects in this book, $1/2$ inch, $5/8$ inch, and $3/4$ inch are

4   Introducing Plywood

the useful thicknesses of softwood plywood. You can get thicker panels, but if you want thinner ones, it is better to choose hardwood.

Hardwood plywood is stronger and heavier than softwood plywood. For many situations you can use thinner panels for the same strength. If plywood is supported by solid wood, the thickness can be as little as 1/8 inch. Although most hardwood plywood comes in 48-inch-×-96-inch sheets, the thickness is quoted in metric measure. There are about 25 mm in 1 inch. This means that 12 mm is just under 1/2 inch, 9 mm is almost exactly 3/8 inch, 8 mm is near 5/16 inch, 6 mm is about 1/4 inch, and 3 mm can be considered 1/8 inch.

One attraction of plywood is its strength in wide and long panels in comparatively thin pieces. Because of this, it is unusual for sheets to be available generally in great thicknesses. You are unlikely to have much use for plywood over 3/4 inch thick. For rather thicker construction, you might use solid wood or another type of board, detailed in the next section.

Quality

You do not just go to the lumberyard and ask for a sheet of 1/2-inch plywood—period! There is much more to it than that. Plywood is now made in a large range of qualities and forms at many prices. There is no point in insisting on the most superior grade when an inexpensive and lesser quality will be just as good for your purpose.

Plies

If you want maximum rigidity in a particular thickness, look at the number of plies. There will not be less than three, as an odd number ensures the grain on both faces being the same way (Fig. 1-1A), but there could be five (Fig. 1-1B) or

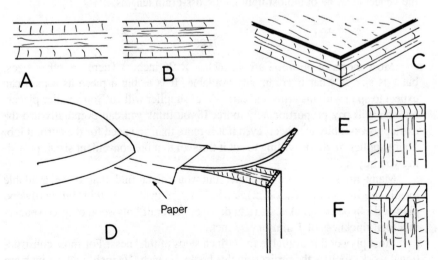

Fig. 1-1. Plywood can have different numbers of veneers and be edged in various ways.

even more. This applies mainly to hardwood plywood. There will probably be only one configuration in softwood plywood.

As the veneers are cut around the log, the knife will shear across knots where there were branches. In softwood, such as Douglas fir, the number of knots depends on the part of the tree trunk being cut. If knots are sound, they cut smoothly and are part of the surface and present little problem. If knots are loose, because they came from an old or dead branch, they are often black around the edges and will fall out.

If all you want is plywood for a temporary barrier or some other unimportant use, you can use the least expensive plywood with empty knot holes and a generally rough appearance. For better work, pick plywood that has been patched. The knot holes are cut out during manufacture and a round or diamond-shaped patch is inserted and finished level. The patch will not show under paint, but it might be rather obvious under a clear finish.

Some plywood is made with a good quality layer on each surface but inferior wood inside. You have to take a chance on the core veneers, but for many purposes it has ample strength, and what is inside does not matter.

Although surface plies will not usually show joints (and if there are any they are close), in the core veneers there might be more than one piece making up the width across a sheet. Ideally, these edges meet, but in much plywood there are gaps or voids, sometimes as much as $1/2$ inch between edges. Obviously, there is no strength in a space, but providing the voids are far apart in different layers, the sheet might be acceptable for most uses. If the edge of such a sheet will show, fill a void with wood putty.

If there is a special wood veneer on a surface, it might be thin and laid either way, but it is marginally stronger if its grain crosses that of the outer plywood veneer (Fig. 1-1C). A *face veneer* might be applied to the other side, even if it will not show, to provide a balance and reduce any risk of warping, which will be slight unless the plywood is quite thin. A face veneer also might be applied to Formica or similar plastic. There could be a decorative piece on top and a balancing plain piece underneath.

Although very thick plywood can be made, the multiplicity of plies makes manufacturing expensive. An alternative is *blockboard* or *solid-core plywood*. In solid-core plywood, there are usually two veneers of the usual plywood type on both surfaces, then strips of solid wood are glued edge-to-edge and between and across the outer veneers.

Glues

An important consideration is the glue used. Earlier plywood suffered from weak glue. Three-ply, as it was known because that was all it was, had a bad reputation because it was joined with the only available natural glues, and it tended to fall apart. Modern glues have considerable strength, but those used in plywood are not all the same.

In ordinary plywood, the glue is strong and has some resistance to dampness. If your plywood will be used outside or anywhere it might get wet, it should have a fully waterproof glue. The common form of this type is sold as *exterior grade*. The glue is waterproof, but other considerations are the same as for ordinary plywood: veneers can be any quality and can have voids inside.

The most superior waterproof type is *marine grade*, intended for boatbuilding. The inner plies are as good as the outer ones, and there are no voids. Marine-grade hardwood plywood is the finest you can get, but you pay a lot more for it.

Surface Quality

Surface quality can vary a lot. Most hardwood plywood has a smooth, sanded surface that will take any finish with little or no preparation. Because of its nature, softwood plywood can have a very rough surface. You can get softwood plywood with one surface much smoother than the other, which suits many applications where the back will not show. One way of providing a better surface during manufacture is called *Medium Density Overlaid (MDO)*, where a resin-treated fiber surface is fused on exterior-grade plywood to give a much smoother finish under paint.

Edges

The edge of plywood, particularly hardwood, can be finished as it is. A rounded and sanded edge, showing all the plies, might be considered a decorative feature under a clear finish. Softwood plywood might be difficult to finish to this standard, particularly if the grain tends to splinter and if there are voids to fill. For many projects an exposed plywood edge is acceptable. In some constructions, the plywood edges are hidden by solid wood frames or other parts. This is always good design, but there are many places where open edges cannot be avoided.

A painted finish can hide exposed edges, but you meet the problem of unequal absorption of the side and end grain of the veneers. Several coats might give an even finish, but early sealing of an edge is a help. You can brush on glue, then sand it smooth before painting. You can buy sealers or use shellac.

There are self-adhesive, heat-sensitive strips that can be used on edges. They are available in woods to match surface veneers. Even if you do not need to match a surface, this edging could be used and painted over. The strips are wider than the plywood thickness they are intended for, so they have to be leveled after fitting. Put a strip in position with paper over it and use a domestic hot iron to press it down (Fig. 1-1D). You then can level the edge with a block plane or sandpaper wrapped round a piece of wood.

Although the iron-on strips have reasonable strength, where the edge is liable to wear or rough use, it is better to put on a solid lip. You might also wish to do this for appearance. Solid wood of a contrasting color framing plywood can be very attractive. Gluing a strip (Fig. 1-1E) of any thickness directly on might hold, but there is little strength where the glue meets the plywood veneer end grain. You will probably have to supplement the glue with brads. You can get a stronger bond by increasing the glue area. Plow a groove in the plywood and make the lip with a tongue to fit (Fig. 1-1F). The lips could be worked on a wide board and each cut off when the tongue has been made.

Joints

Plywood is unsuitable for cutting many of the joints used in solid wood. Where plywood supplements solid wood, it is often in the form of panels that fit into grooves. Where plywood is the main material, however, you have to join parts in other ways.

You can use nails quite successfully. If you use pins or such types as finishing nails or brads with their small heads, you can set them below the surface and cover them with a stopping or wood putty so that they do not show under a finish. For positions where visible heads do not matter, use box nails.

You can increase the strength of nailed joints by driving at alternate angles in a dovetail manner (Fig. 1-2A). At an end or open top, where more load might be expected, put two or more nails closer together (Fig. 1-2B).

In many assemblies, solid wood might back up plywood panels, either for stiffness or to give a better glue or nailing area in a corner. The solid wood need not be of very large section. It can be glued to the plywood, but if you want to avoid locating and clamping problems, use a few brads from the plywood into the strips.

There is no need to cut such joints as bridles or half laps between the solid wood parts. You can merely overlap the strips (Fig. 1-2C), or at a corner where joints will show, you can miter them (Fig. 1-2D). Stiffening in this way on one side is satisfactory if the assembly will be built into more construction. If the

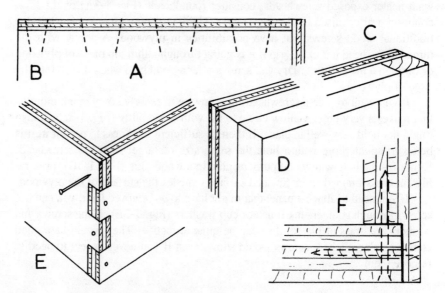

Fig. 1-2. Nailing in different directions strengthens joints.

8   Introducing Plywood

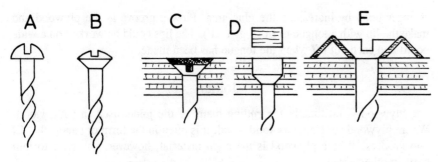

Fig. 1-3. Screws for plywood can be used in several ways.

panel is without other support, as in a door, it is better to have plywood on each side of the strips. If strips are just on one side of one piece of plywood, the alteration of bulk (because of varying moisture content due to changes in the atmosphere) might cause slight warping on a thin panel.

Plywood edges can be nailed together in the same way as solid wood, but the nail grip might not always be very good. It generally is better to arrange joints that permit glue and nails both ways.

Finger joints with wide parts can be cut in plywood (Fig. 1-2E). Do not attempt to cut these joints with narrow parts, as would be usual in solid wood. When nailed both ways, the joint cannot be opened.

Another way of nailing both ways is to cut a rabbet in one piece so the other fits in (Fig. 1-2F). Besides other corners, a rabbet suits a drawer, where you do not want joint details showing at the front.

There are many places where it is better to use screws than nails. Maybe all you want are flathead screws that finish level and not matter if they show. If you want neater exposed screwheads, consider roundheads (Fig. 1-3A) or the less-common oval or raised heads (Fig. 1-3B). There are some other screws besides traditional wood screws that have possibilities in plywood. A screw with the thread to the head has extra grip if it is going through a thin top piece of plywood or through a metal fitting. Drywall screws offer a good grip when driven into the edge of plywood.

If you want to hide a screwhead, it is unwise to merely countersink more as that makes a very large shallow circle to fill with wood putty (Fig. 1-3C). It also might not hold very well in place. If there is sufficient thickness to permit it, it is better to counterbore with a hole the same size or larger than the screwhead. Either fill the hole with wood putty or glue in a wood plug (Fig. 1-3D), preferably cut cross-grained so its finished surface matches the surface of the plywood.

If you need to attach a panel that might have to be removed later, you can use screws through countersunk finish or cup washers (Fig. 1-3E). If the screws and washers are brass or plated, they can be quite attractive. They would be a good choice in places where screws would show, even if you never expect to need to remove them.

2

Tables and Stands

Take-down Coffee Table

The take-down coffee table in Fig. 2-1 is a small octagonal table that could be made as a permanent assembly, but it is designed to separate into three pieces that can store flat. The two leg pieces notch into each other and fit into strips under the top.

Make the top from 1/2-inch plywood and the legs from 3/4-inch plywood. If you use softwood plywood, you might paint the table. The top can be veneered with a choice hardwood or covered with a melamine plastic to protect it from heat and liquids. You can then paint the legs, although clear varnish on hardwood plywood also will look good.

Fig. 2-1. This octagonal coffee table can be taken apart to pack flat.

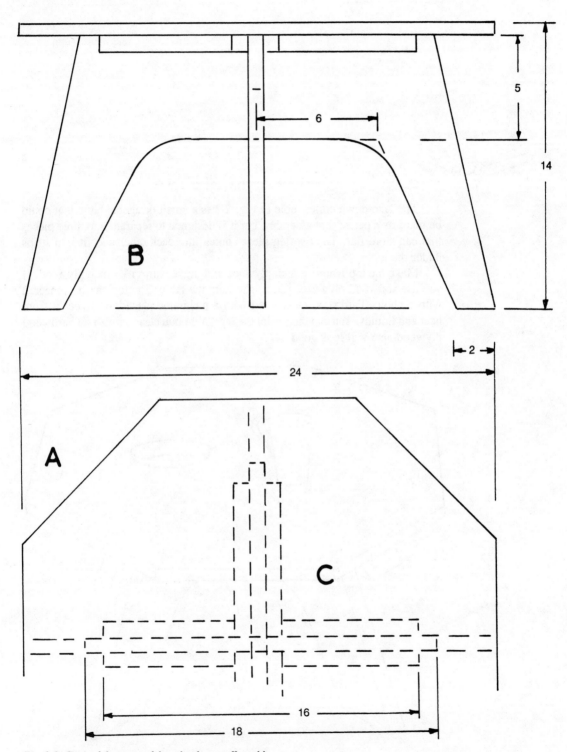

Fig. 2-2. Sizes of the parts of the take-down coffee table.

1. Mark out the top (Figs. 2-2A and 2-3A) first as a square. Divide its underside centrally across the leg positions. If you measure the diagonal distance from the corner to the center and mark this distance from the corners along each side, you will get the points from which to mark the octagon. Cut the outline.
2. Mark out a pair of legs (Fig. 2-2B). The feet extend as far as the top and slope to 3 inches in. Cut the outline and mark the second leg from it.

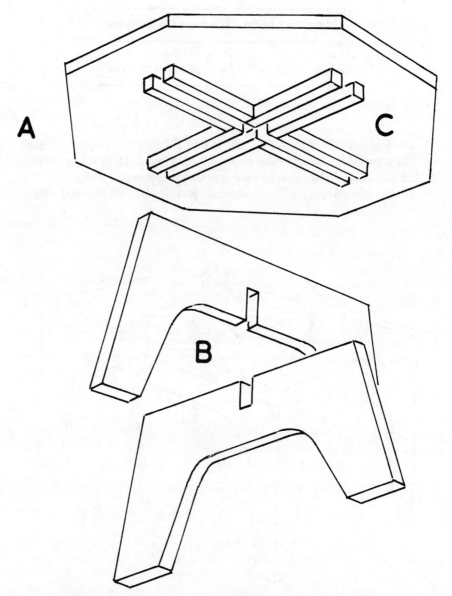

Fig. 2-3. Method of assembly of the take-down coffee table.

3. Notch the centers of the legs pieces to fit each other (Fig. 2-3B).
4. Mark the thickness of the legs on the underside of the top. Fit strips across outside these marks (Figs. 2-2C and 2-3C) so the assembled legs can be pressed in. The outer ends of the strips are 1 inch from the outsides of the legs.
5. Round the outer edges of all parts.
6. When you are satisfied with the fit of parts and their ability to disassemble, apply your chosen finish.

Materials List for Take-down Coffee Table	
1 top	24 × 24 × 1/2 plywood
2 legs	14 × 24 × 3/4 plywood

Coffee Table

The legs and rails of the coffee table in Fig. 2-4 are cut from plywood panels. The plywood top is surrounded by a solid wood frame. There is ample rigidity, but the table is lighter than its solid appearance implies.

For the sizes shown (Fig. 2-5A), the main parts are 1/2-inch plywood with 1-

Fig. 2-4. This coffee table has a framed top and stiffened plywood legs.

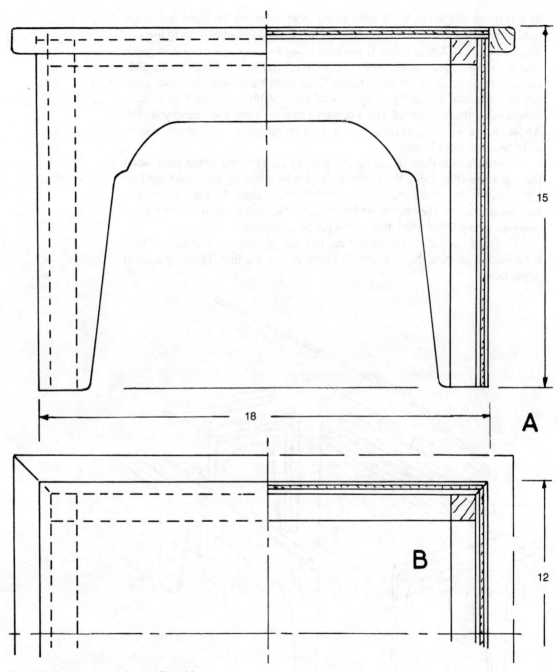

Fig. 2-5. Suggested sizes for a coffee table.

inch nominal strips inside. A table made from softwood plywood and strips would be lightest and would look attractive if painted in bright colors to use in a playroom, den, or family room. If you want to use the table in a room with good-quality furniture, you can use hardwood plywood and a clear finish, so the table can take its place alongside other things made from hardwoods. In either case, you could feature the top by using a faced piece of plywood, such as a store-bought panel already covered with a veneered pattern or one you veneer yourself. An alternative would be to cover the top with melamine or other plastic able to resist heat and most liquids.

Construction is done by gluing and pinning the plywood to the solid wood framing strips (Fig. 2-6). You could avoid pins by clamping each joint until its glue sets, but that would make total assembly rather slow. Pins serve the same purpose as clamps. They can be set below the surface and covered with stopping, although for a painted finish that would not be necessary.

The same method of construction can be used for tables of other sizes. There is no need to increase the plywood thickness, unless you intend making a much larger table.

Fig. 2-6. How the parts of the coffee table are arranged.

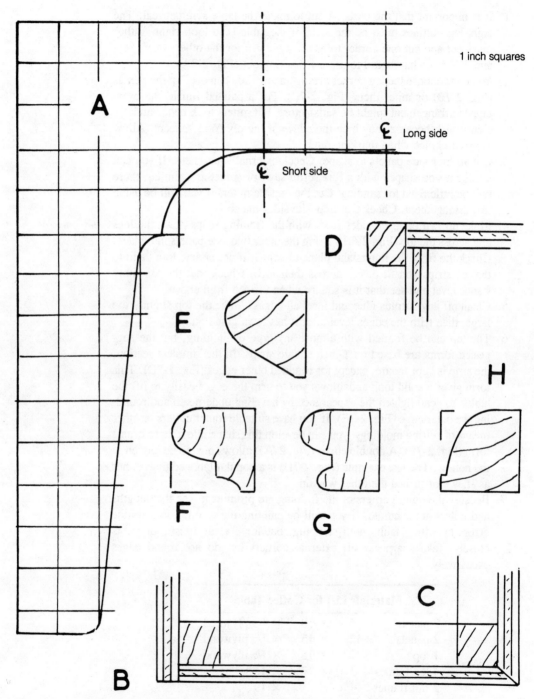

Fig. 2-7. Leg shapes and border sections for the coffee table.

1. It is important that the parts be cut to make the table symmetrical. The eight leg outlines must be the same if the table is to look right. Either mark out and cut one corner to use as a pattern for the others or make a template from hardboard to use for marking out (Fig. 2-7A).
2. You must choose the leg corner arrangements. You can overlap the panels (Fig. 2-7B) or miter them (Fig. 2-7C). For a painted finish, the overlapped arrangement might be satisfactory, but miters look better under a clear finish because they hide the edges of the plywood veneers. Allow for your choice when marking out.
3. Cut all four side panels to shape. Check external squareness. If you can cut the inside shapes with a fine scroll saw, or something similar, there will be little need for sanding. Cut the outside miters if you will be using that arrangement. Check that opposite sides match.
4. Make up two opposite sides first, with the framing strips down the legs and across the top (Fig. 2-5B). Bring the other two side panels into place. Check the fit of the miters and plane to adjust, if necessary. Join them to the leg strips and fit pieces across their tops. Check that the assembly stands level and see that it is square when viewed from above.
5. Clean off any surplus glue and level all edges. Make the top slightly too large, then trim the edges level after it has been glued on.
6. The top can be framed with almost any type of molding, but the suggested forms are based on 1-inch square strips. In the simplest section, the strip is kept square, except for rounded outer edges (Fig. 2-7D). This form gives a solid look and allows you to grip the edges easily to lift the table. You can lighten the appearance by beveling underneath and rounding the outer edge (Fig. 2-7E). If you have suitable router cutters or other means of cutting moldings, you can develop this shape into a more ornate form (Fig. 2-7F). A double curve (Fig. 2-7G) allows pins to be hidden in the hollow. The last example (Fig. 2-7H) is a popular table edging, either as shown or turned the other way up.
7. Because the outer corners of the framing are prominent, see that lengths and miters are accurate. If you will be painting the wood, you can hide errors by filling faults with stopping, but for a clear finish, fit joints closely. Take sharpness off external corners but do not round edges excessively.

Materials List for Coffee Table			
2 panels	15	× 18	× 1/2 plywood
2 panels	12	× 15	× 1/2 plywood
1 top	12	× 18	× 1/2 plywood
4 leg frames	7/8	× 7/8	× 16
2 rail frames	7/8	× 7/8	× 19
2 rail frames	7/8	× 7/8	× 13
2 top frames	1	× 1	× 22
2 top frames	1	× 1	× 16

Pedestal

The pedestal in Fig. 2-8 can serve as a stand for a vase of flowers or a potted plant. It is also suitable for holding a reading lamp beside a chair or bed. It has a broad spread of feet for stability. If you alter sizes, make sure the feet spread several inches more than the size of the top.

Construction is completely of 1/2-inch plywood, except for a central 1 1/2-inch square pillar. The legs fit around the pillar (Figs. 2-9 and 2-10A) and so are offset, but this simplifies construction and adds interest to the design. Softwood plywood could be used if you choose a painted finish, though hardwood plywood with a clear finish might look better with other furniture. If you anticipate water spilling onto the top, make it of plywood with a plastic facing.

1. Make the four legs (Figs. 2-10B and 2-11A). Because it is important that the legs match, make one and use it as a pattern to mark the others, or make a hardboard template for marking all four.
2. Sand and lightly round the exposed edges.
3. Mark the positions of the shelves (Figs. 2-10C and 2-11B). They alternate, with one opposite pair higher than the other pair, for appearance and so pins or screws can be driven into their edges.
4. Make the pillar. It must be parallel with all its corners at 90-degree angles and widths the same both ways if the pedestal is to assemble squarely.
5. Attach the legs to the pillar with glue and pins (Fig. 2-10D). Make sure each straight edge beds tightly against the next leg. Check that ends are level with each other.
6. The four shelves are the same, but because of the offset legs, they are not parts of regular octagons. Cut them to shape and round their outer edges (Fig. 2-11C).
7. Fit the shelves in the marked positions, using glue and pins or fine screws.
8. The top is a regular octagon (Fig. 2-11D). It is shown with square edges. You cannot cut moldings on plywood edges satisfactorily, but you can round them or fit solid wood or plastic molded strips.
9. The top has to be centered (Fig. 2-10E), but because of the offset arrangements of the legs, you cannot use them as a guide. Drill a hole in the center of the top and another in the center of the pillar (Fig. 2-10F). Counterbore the hole in the top, so it can be filled with a plug or stopping. Glue on the top and secure it with the screw. You might need to drive a pin near the outer end of each leg when these have been arranged symmetrically.
10. Before applying a finish, check that the pedestal stands upright without rocking. If necessary, plane the bottoms of one or more legs.

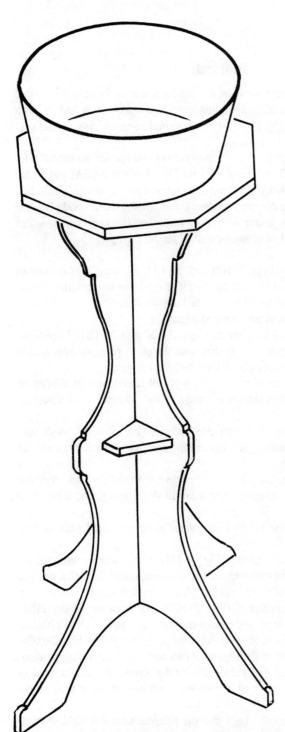

Fig. 2-8. The pedestal stands at table height and will support a plant or other decoration.

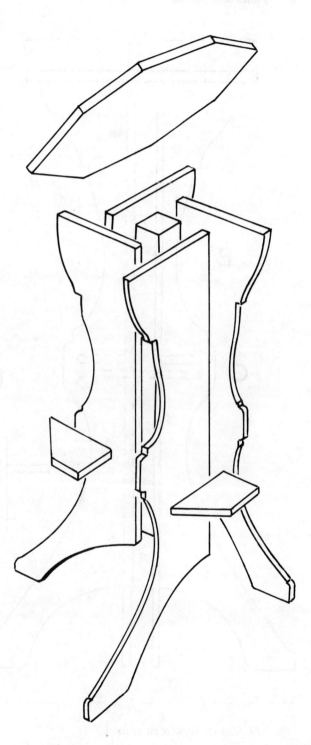

Fig. 2-9. The pedestal has legs around a central post.

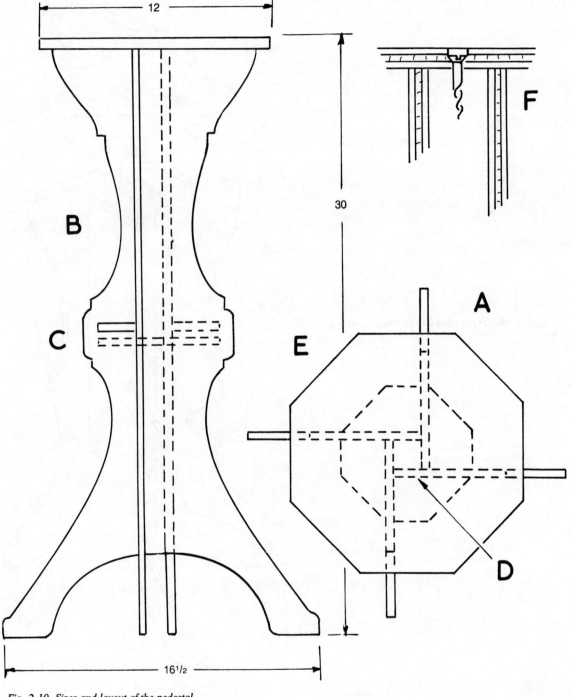

Fig. 2-10. Sizes and layout of the pedestal.

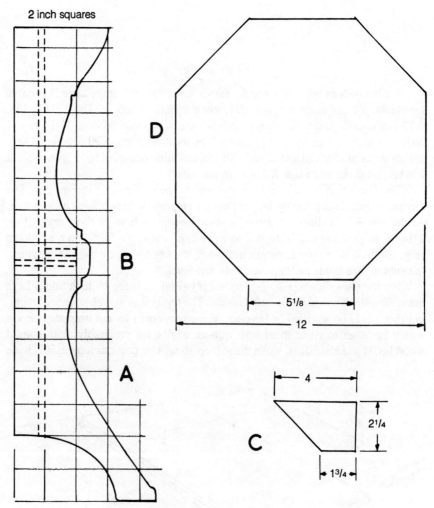

Fig. 2-11. Outlines and parts of the pedestal.

Materials List for Pedestal			
4 legs	10	× 30	× 1/2 plywood
1 top	12	× 12	× 1/2 plywood
4 shelves	3	× 5	× 1/2 plywood
1 pillar	1 1/2 ×	1 1/2 ×	27

Flap Table

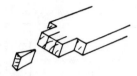

A table with an adjustable top for size is always useful. If the table is suitable for meals, you can adapt it to seat different numbers of people. The table in Fig. 2-12 is a suitable height for meals, and its top can be changed from 18 inches × 36 inches to 36 inches square by turning it on its stand through 90 degrees. There are no struts or extra legs to adjust. The folded table occupies little space, but it can be pulled out and made full size in seconds.

The lower part is made in a similar way to the coffee table in Fig. 2-4. The top has a central part slightly bigger than its support, with two flaps hinged to it. At the center is a disc that forms a pivot through a hole in the subtop (Fig. 2-14A), so the table can be turned to a position where the flaps can hang down (Fig. 2-13A). When the flaps are lifted and the top turned through 90 degrees, the subtop then holds the flaps securely and level.

For the sizes shown (Fig. 2-13B), all plywood parts are 3/4 inch thick. Legs are stiffened with solid wood strips inside. The top is shown with square corners, but they could be rounded or beveled. The edges could be left untreated, but it would be better to cover them with iron-on edging or, preferably, with a solid wood lip. If you alter sizes, make the subtop about 1 1/2 times as long as it is wide

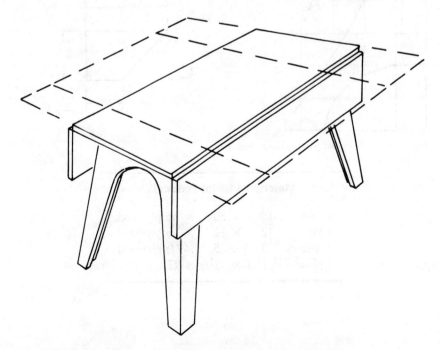

Fig. 2-12. This flap table has a rotating top. The frame supports the flaps.

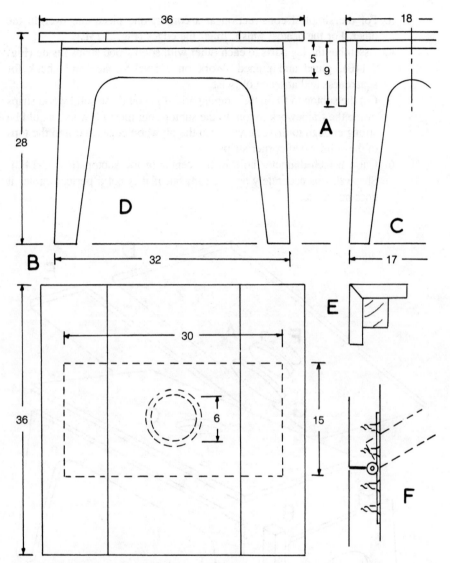

Fig. 2-13. Sizes and layout of the flap table.

and the center part of the top wide enough to allow the flaps to hang. The length of the top the other way is not so important.

1. Set out and cut the end leg shapes (Fig. 2-13C). Set out the slight outside taper by making the overall width 2 inches more at the bottom than at the top. Taper the legs from 3 inches at the bottom to 5 inches at the top and make the horizontal part 5 inches deep. Cut a curve between the legs.
2. Use the end legs as templates to mark out the leg shapes and angles on the longer sides (Fig. 2-13D).

3. For the simplest construction at a corner, one piece can overlap the other. For the neatest finish, miter the edges (Fig. 2-13E).
4. Assemble the leg parts to each other with solid wood strips inside (Fig. 2-14B), glued and pinned. Work on a level surface and check for squareness and absence of twist.
5. Cut the subtop to fit on this framework. You could use solid wood strips inside the framework to join to the subtop, but most plywood should be strong enough to drive screws into the plywood edges and into the tops of the solid wood corner strips.
6. Cut a 6-inch diameter hole in the center of the subtop (Fig. 2-14C). Shape this as accurately as you can, but if it is not a perfect circle, it does not matter.

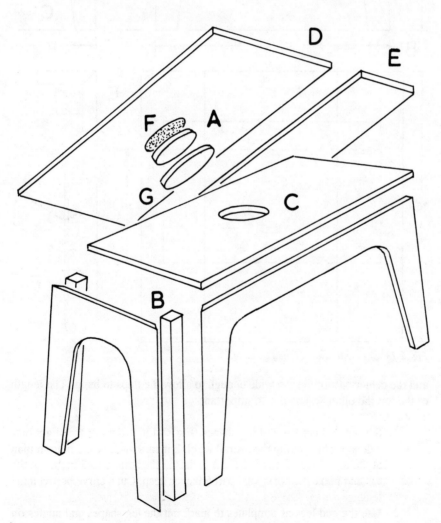

Fig. 2-14. Parts of the flap table showing the rotating arrangements.

7. Make a disc the same thickness to revolve easily in the hole (Fig. 2-14A). If you have cut the hole carefully, you might be able to use the waste piece from it. Slight play in the fit does not matter.
8. Cut the parts of the top. Make sure the meeting edges are straight. The center piece is 18 inches wide and the two flaps are 9 inches wide, although they could be a few inches wider if you want a longer table when opened. Round or bevel the outer corners or add lipping, if you wish.
9. Attach the disc at the center of the top. To increase thickness and provide clearance, include a circle of stout paper in the joint (Fig. 2-14F).
10. Make an 8-inch disc as a retainer (Fig. 2-14G). Screw it in place on the other disc without glue and test the rotating action. Remove the retaining disc.
11. Join the two flaps with back-flap hinges (which swing back further than ordinary hinges). Three 1 1/2-inch-wide hinges should be suitable (Fig. 2-13F). Let them in so nothing projects to interfere with rotating over the subtop.
12. Finish the wood with varnish, polish, or paint before fitting the tabletop on the subtop and replacing the retaining disc.
13. Casters can be fitted to the bottoms of the solid wood strips inside the plywood legs.

Materials List for Flap Table

2 end leg frames	17	× 28	× 3/4 plywood
2 side leg frames	28	× 33	× 3/4 plywood
1 subtop	15	× 30	× 3/4 plywood
1 top	18	× 36	× 3/4 plywood
2 flaps	9	× 36	× 3/4 plywood
1 disc	7	× 7	× 3/4 plywood
1 disc	9	× 9	× 3/4 plywood
4 legs	1 1/2	× 1 1/2	× 28

Folding Desk

A desk or work top is a useful piece of furniture for dealing with home accounts or doing school work, sewing, or a hobby, but when it is not in use, it might take up too much room. The desk in Fig. 2-15 has a pedestal cabinet with storage space and a long kneehole working area that can be folded with its leg against the side of the cabinet when not in use.

When opened for use, the top is 18 inches wide × 51 inches. When folded, the cabinet top area is 18 inches × 22 inches, with all the lower parts within that area. Height in both cases is 30 inches. The cabinet is described with an open shelf above a cupboard, but you could equip it with drawers or in any other way to suit tools or whatever you wish to store.

28 Tables and Stands

Fig. 2-15. The flap and leg of this desk fold against the cabinet.

Most parts are 3/4-inch plywood. The back of the cabinet is hardboard or thin plywood. If the desk is to form part of the furnishings of a living room, the plywood for the top and exposed parts would look best if veneered with attractive wood or plastic, with matching edging ironed on. A wood lip on the top would be better able to keep its appearance after long use.

Joints could be glued and screwed, with the heads counterbored and plugged, or you could use 5/16-inch dowels at about 3-inch intervals. If you alter sizes, the length of the flap and its leg have to be related to the height of the cabinet, as will be seen in instructions #7 to #9. The cabinet is of straightforward construction and should be completed first. The depth is 17 inches from front to back, including the hardboard. The top overhangs 1 inch at one side and at the front, and extends 3 inches at the side where the flap will hinge (Fig. 2-16A), which could be on either side.

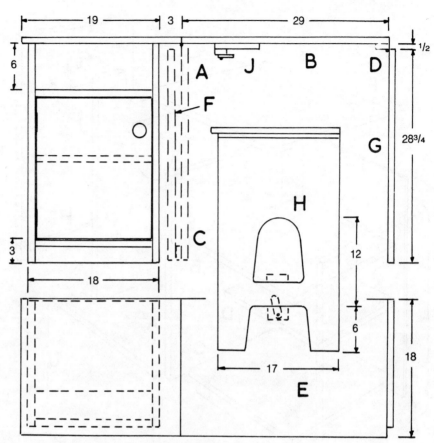

Fig. 2-16. Suggested sizes for the folding desk.

1. Make the two cabinet sides (Fig. 2-17A). Mark on the positions of other parts.
2. Cut the pieces that go crosswise to the same length. The shelf (Fig. 2-17B) and the cupboard bottom (Fig. 2-17C) are the same width as the sides. The middle shelf can be 12 inches wide (Fig. 2-17D). The toe board is set back a little (Fig. 2-17E).
3. Mark dowel positions on the cabinet pieces or prepare them for screwing. Assemble with glue between meeting surfaces.
4. Prepare the back hardboard with enough at the top to go over the cabinet top later. Glue and screw it in position to keep the assembly square.
5. Prepare the cabinet top (Fig. 2-17F) with any surface veneer and edging. Fix it to the sides with dowels. Screw the hardboard to it.
6. The door (Fig. 2-17G) should be cut to fit easily in the opening. Use two 2-inch hinges at one side. Put a knob or handle at the other side, high enough to be reached comfortably. Fit a spring or magnetic catch, or you might prefer a lock.

30 Tables and Stands

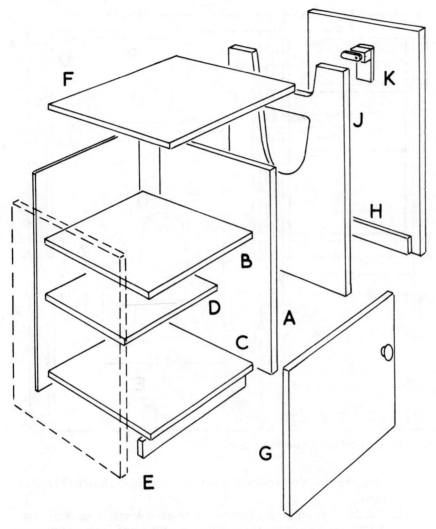

Fig. 2-17. Parts of the folding desk.

7. Make the length of the flap ¹/₄ inch less than the cabinet height under its top (Fig. 2-16B), so it can swing down without rubbing the floor (Fig. 2-16C). It is a simple rectangular piece finished to match the cabinet top.
8. To give clearance for the leg to fold, it is necessary to thicken the outer end of the flap. A strip about 2 inches wide and ¹/₂ inch thick will do (Figs. 2-16D and 2-17H) and can be plywood or solid wood.
9. The leg is the same width as the cabinet. Its length must be sufficient to hold the flap level, yet it has to fold inside the flap (Fig. 2-16F). The ¹/₂-inch packing on the flap allows the length to be ¹/₄ inch less than the length of the flap (Fig. 2-16G).

10. Cut away the leg to form two feet (Fig. 2-16H). The exact shape cutout is not important, but include a crossbar (Fig. 2-17J).
11. When the desk is folded, the leg should be held to the flap, then gravity will ensure that it hangs vertically without fastening. Put a 1/2-inch strip pad under the flap to come behind the folded leg and extend 2 inches for a 3/4-inch thickening piece (Figs. 2-16J and 2-17K). Arrange the pad so you can use a thin wood turnbutton to secure the leg when the leg hinges under the flap.
12. Hinge the flap to the cabinet top and the leg to the thickening piece on the flap with two 2-inch hinges at each place. Test the folding action, then you might wish to remove the hinges while you finish the wood. You could use a dark stain, followed by varnish or polish. You also could paint the lower parts and put a clear finish on the top, if it is attractively veneered.
13. You might find the leg stands upright without further support, but you could attach a folding metal strut to the rear edges of the flap and leg.

Materials List for Folding Desk

2 sides	17	× 30 × 3/4 plywood
2 shelves	17	× 18 × 3/4 plywood
1 shelf	12	× 18 × 3/4 plywood
1 toe board	2	× 17 × 3/4 plywood
1 door	17	× 19 × 3/4 plywood
1 top	18	× 22 × 3/4 plywood
1 back	18	× 31 × 1/8 hardboard
1 flap	18	× 30 × 3/4 plywood
1 leg	17	× 30 × 3/4 plywood
1 strip	1/2	× 2 × 18
1 pad	1/2	× 3 × 7
1 pad	1/4	× 2 × 4

Greenhouse Plant Stand

In most greenhouses, every bit of space has to be used. Tiers of shelves, like the stand in Fig. 2-18, will employ space to the maximum. The stand has three shelves and space for things like seed trays underneath. It is made in sections, so you can either build it permanently or take it apart when not needed to clear the area. Its construction also allows you to assemble the stand in place if the door is not large enough for it to pass through when complete. Besides using the stand in a greenhouse, you could use it to display your plants or flowers on a patio or deck or to display your wares at a fair.

The sizes suggested (Fig. 2-19A) include shelves that are 11 inches wide with a small overlap. They are 12 inches apart vertically, so the whole stand is 36

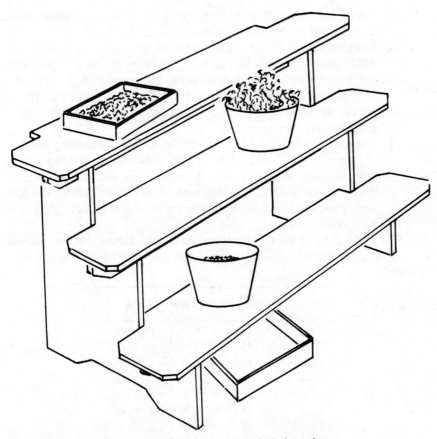

Fig. 2-18. The greenhouse plant stand is built with notched plywood parts.

inches high and 26 inches wide, with whatever length you need—in this example it is 48 inches. If you want a very long stand, it might be best to make two parts, for convenience in moving, or you could make it in one piece with an intermediate upright.

All parts could be ¾-inch exterior plywood, although the shelves might be reduced to ½ inch if you are not intending to have excessive loads. The rails attach to the shelves, which are interchangeable. The rails fit into slots in the uprights, and the rear 2 inches of the lower shelves fit alongside the uprights for additional strength and to resist twisting. If you lift out the shelves with their attached rails, the parts will pack flat. If you do not want the take-down option, glue the slotted parts together. As conditions might be damp, use waterproof glue.

1. Make the two sides (Figs. 2-19B and 2-20A). Cut the notches to suit the plywood for the rails. Cut away the bottom edge to form feet.
2. The three rails control the length of the assembled stand (Figs. 2-19C and 2-20B). Notch the ends to fit the uprights and bevel the lower corners.

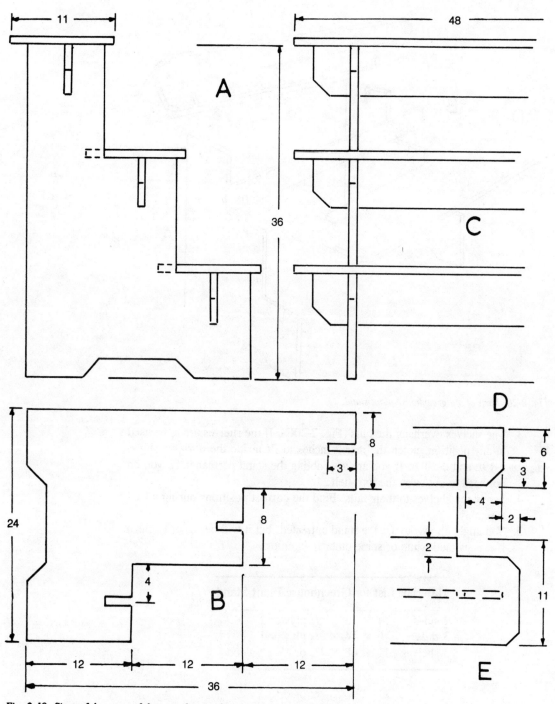

Fig. 2-19. Sizes of the parts of the greenhouse plant stand.

34 Tables and Stands

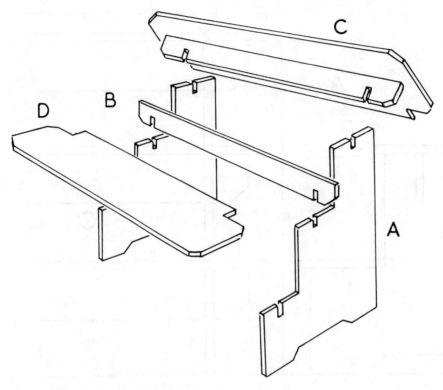

Fig. 2-20. Parts of the greenhouse plant stand.

3. The shelves overhang the rails (Fig. 2-20C). If the shelves are to be used in any position, notch the rear 2 inches to fit inside the uprights (Figs. 2-19E and 2-20D). If you are assembling the stand permanently, you do not need to cut away the top shelf.
4. Screw the shelves to their rails. Find the correct positions during a trial assembly.
5. You might choose to use the stand untreated, but it will be easier to clean if it is painted white or some other light color.

Materials List for Greenhouse Plant Stand

2 sides	24 × 36 × 3/4	plywood
3 rails	6 × 44 × 3/4	plywood
3 shelves	11 × 48 × 1/2 or 3/4	plywood

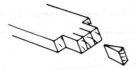

Bed Table

If meals have to be served in bed—to an invalid, for example, or just because you enjoy the luxury of it—you need a table to enjoy the food in comfort and keep it off the bed covering. A tray on the knees is unsteady. The table in Fig. 2-21 is like a large tray with legs that stand on the bed on each side of the user. When not in use, the legs fold flat under the tray. There is nothing to adjust or lock as the legs are kept in the open or closed positions by a wood spring.

The suggested sizes (Fig. 2-21A) are for a tray 12 inches wide and 26 inches long, standing 10 inches high. If you alter the sizes, make sure the legs fold without overlapping. The table has a border on the top around three sides (Fig. 2-21B). The legs are hinged underneath (Fig. 2-21C). A springy piece of wood on a block extends over the centers of the legs (Fig. 2-21D) and holds the legs out (Fig. 2-21E and F) or closes over them when folded (Fig. 2-21G and H).

All parts except the spring are $1/2$-inch plywood. The spring could be $1/4$-inch hardwood plywood, but it might be better to use solid wood. Any wood with a moderate spring will do. Ash or hickory is ideal, but not much spring is needed and many woods are suitable. Experiment with what you have.

1. Cut the tray base to size. Make the border (Fig. 2-21B). A height of $1 1/2$ inches should be satisfactory. Glue and screw from below. Miter the rear corners, slope down the ends, and round the top edges.
2. Make the two pairs of legs. Cut the top and bottom to 80 degrees, so the legs splay when opened. Cut the inner edges square. Where the spring comes will be beveled later. Round all edges and the bottom corners of the legs.
3. Put two 2-inch hinges on each leg (Fig. 2-21C) so that when the legs are flat, the hinges are closed; then when pulled open, the 10-degree leg top fits under the tray and the leg splays correctly. Screw the hinges temporarily to the underside of the tray.
4. Make the spring (Fig. 2-21D) almost as long as the tray. Its section depends on its springiness, but $1/4$ inch $\times$ $1 1/2$ inches will probably be satisfactory. Round its edges and ends. Mount it by screwing to a block 4 inches long and $1/2$ inch thick at the center of the underside of the tray.
5. The spring extends over the center bars of the legs. Try its effect when you lift a pair of legs. Make a notch at the angle of the spring in each bar (Fig. 2-21E). If the spring is too stiff, thin the wood or deepen the notch in each leg bar. When both pairs of legs are opened, the table should stand rigidly on a bed or other surface.
6. How you finish the table depends on the wood used and your preference, but do not make the top surface slippery.

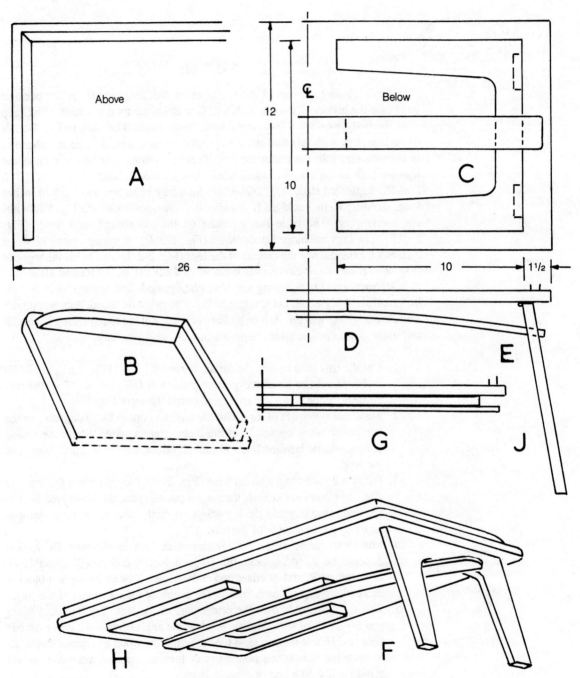

Fig. 2-21. This table can stand on a bed as a tray. It has legs that fold under the tray and are held by a wood spring in either position.

Materials List for Bed Table

1 tray	12	× 26	× 1/2 plywood
2 legs	10	× 10	× 1/2 plywood
1 border	1 1/2	× 26	× 1/2 plywood
2 borders	1 1/2	× 12	× 1/2 plywood
1 block	1 1/2	× 4	× 1/2 plywood
1 spring	1/4	× 1 1/2	× 26

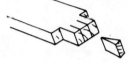

Game Table

This is a table that is normally 26 inches square, but it can be opened to about 36 inches square. The height is 30 inches (Fig. 2-22). The table could have many uses but is ideal for a game table. The top is in two thicknesses, with the upper part divided diagonally. The folded triangles show one surface. When the triangles are opened out, their other sides extend the top, which was previously hidden underneath. You could put baize or other cloth on one surface for card games. You might paint a surface as a chessboard or board for other games.

Except for leg stiffening, all parts can be 3/4-inch plywood. There are four pullout slides to support the opened flaps, and they run on guides that could be made completely of plywood or of a plywood base with solid wood strips. You could build the table with any plywood for a painted finish, but the top parts would look better if made of faced hardwood plywood and given a clear finish.

It will be best to start by making and assembling the legs, then adding the slide and guides, then finally putting on the tops. Variations in size will not matter, but the instructions apply to the suggested sizes in Fig. 2-23A.

1. The four leg parts are the same, if you miter the corners (Fig. 2-23B). It would be simpler to overlap them, but you would then have to allow for the different widths.
2. Mark out the legs (Fig. 2-24A). Curves between the legs and rail provide stiffness. Cut the slots to easily pass 3/4-inch plywood.
3. Join the leg corners with 1 1/2-inch square strips inside the joints (Fig. 2-23C). Check for squareness and lack of twist. If necessary, put a temporary strip across the top to hold it in shape.
4. The slide guide is a plywood cross (Fig. 2-24B) fitted between the rails and level with the bottoms of the slots. Make each arm 3 1/2 inches wide (Fig. 2-24C). Round the corners between the arms.
5. Put plywood or solid wood strips on this plywood base, spaced to allow the 3/4-inch plywood slides to move easily (Fig. 2-24D). The strips can go right across in one direction, while the others butt against them. Put a stop block at the center (Fig. 2-24E).
6. Fit the slide assembly on blocks under the rail slots (Fig. 2-23D).
7. Make four slides (Figs. 2-23E and 2-24F). Arrange the lengths so that

38 Tables and Stands

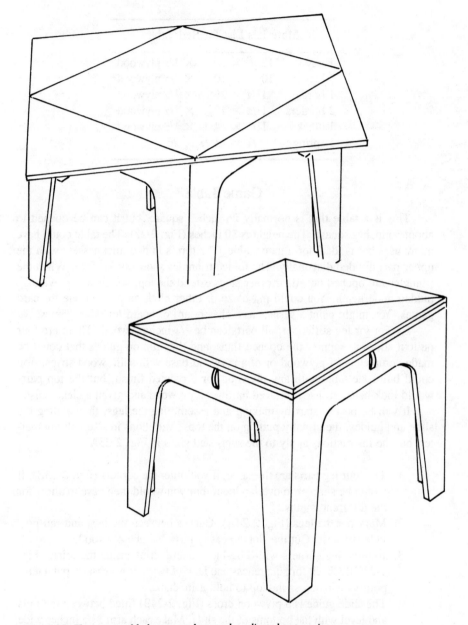

Fig. 2-22. This game table has a top that can be adjusted to two sizes.

when a slide reaches the center stop, the outer end can have a curved end projecting far enough to provide a finger grip. The depth should allow easy movement under the top, but avoid slackness, which might allow a flap to sag.

8. Put stops on the sides of each slide (Fig. 2-24G) to allow about half the length of the slide to project when pulled out.

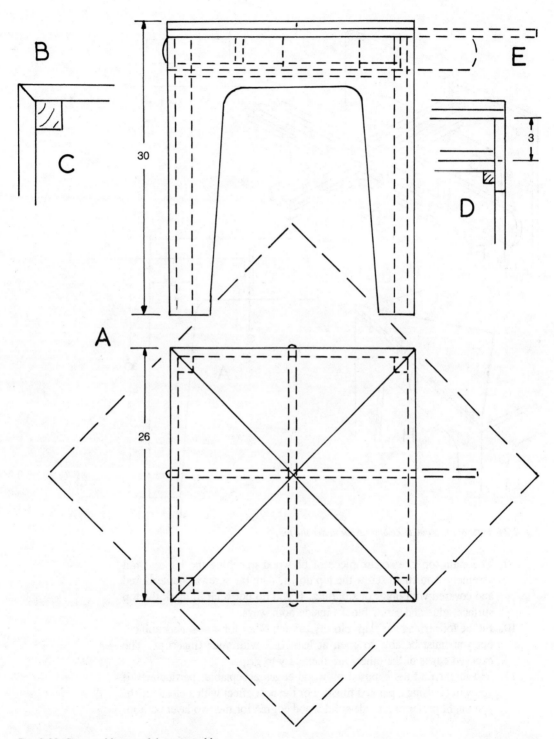

Fig. 2-23. Sizes and layout of the game table.

40 Tables and Stands

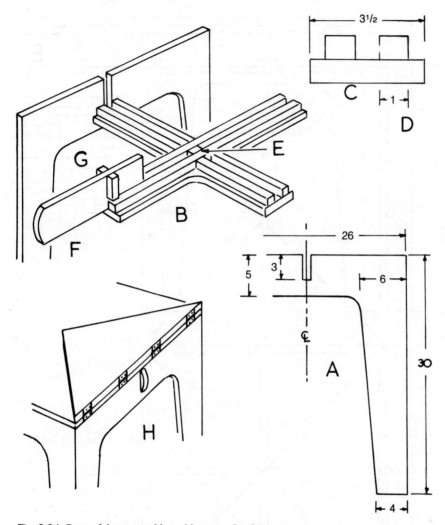

Fig. 2-24. Parts of the game table and layout under the top.

9. The main top is a square piece of plywood matching the leg and rail assembly. You could screw the top down, with the screws counterbored and covered with plugs or stopping. If you do not want to mark the top surface, glue and screw blocks inside both ways.
10. Fit the four triangular flaps closely to each other for a neat appearance, but you must be able to open at least one with your fingertips. The exposed edges of the others are then easy to grip.
11. You might find the exposed plywood edges acceptable, particularly if you will be using a painted finish. For the best effect with a clear finish, you might prefer to provide solid wood edging for the two layers of top.

12. Hinge the edges (Fig. 2-24H). Because the hinges show, they will look best if they and the screws are brass or plated. If possible, use hinges that cover the double thickness of plywood. This looks better than narrower ones. Width is more important than hinge length, which will probably be about 2 inches.
13. Try the action. Remove sharp edges and apply your chosen finish.

Materials List for Game Table

4 pairs of legs	26	× 30	× ³/₄ plywood
1 guide	25	× 25	× ³/₄ plywood
4 slides	3	× 15	× ³/₄ plywood
1 top	26	× 26	× ³/₄ plywood
4 flaps	13	× 26	× ³/₄ plywood
4 legs	1¹/₂ ×	1¹/₂ ×	30
8 stops	³/₄ ×	³/₄ ×	4
4 guides	³/₄ ×	1	× 25

Corner Table

In many rooms, a corner is a wasted space. If a table can go right into the angle, it provides useful space that might not otherwise be used. A simple triangle does not give much top area, but by squaring the corners, a table 24 inches along the wall has space large enough for such objects as a television set plus ample storage space underneath (Fig. 2-25).

This table can be made entirely of ³/₄-inch plywood, but you could reduce thickness to ¹/₂ inch for some of the internal parts. You might use plywood with an attractive surface veneer for the parts that show. The exposed edges could be lipped with solid wood, matching the veneer for the best appearance.

It is unwise to assume that the corner of a room is square. It might be, but test it with a large square at the top level and make the tabletop to match the actual shape, then fit the other parts to it. Besides being out of square, a wall might not be straight. If you make the table corner square when the room corner is not, you will finish with unsightly gaps. If there is a baseboard, you might cut around it or remove it where the table comes. It might be possible to make the height of the table toe board the same as the wall baseboard, then only that will have to be cut back.

Because of the triangular shape, some construction might not be immediately obvious. The plan view in Fig. 2-26A is the important one for settling layout. The view looking into the corner (Fig. 2-26B) shows what will be seen from the room. The view along a wall (Fig. 2-26C) shows that only the top goes to a point. The internal parts stop 12 inches along the wall (Fig. 2-26D).

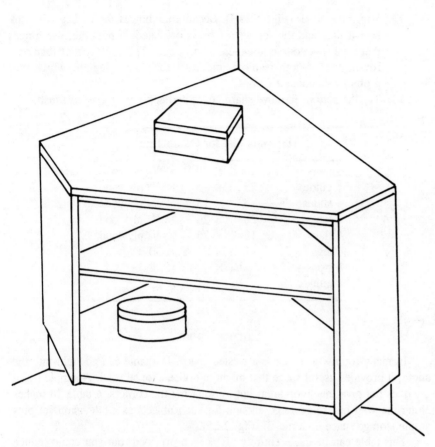

Fig. 2-25. This table with shelves makes good use of the corner of a room.

1. Mark out the top (Figs. 2-26A and 2-27A). Cut it to fit the corner at the level it will be. Mark on its underside the positions of other parts. The top has 12-inch sides (Figs. 2-26E and 2-27B) and a subtop (Fig. 2-27C) against the sides and behind the uprights (Figs. 2-26F and 2-27D).
2. Make the 6-inch wide uprights. The uprights are parallel and reach from under the top to 4 inches from the floor.
3. Make the two 12-inch wide pieces that go against the wall and behind the uprights (Fig. 2-26G). They reach the floor.
4. Shape the bottom to fit against the wall pieces and the uprights. The front edge can come level with the inner edges of the uprights. The rear edge comes inside the wall pieces.
5. The subtop (Fig. 2-27C) has the same front outline as the bottom, but is 6 inches wide.
6. Make the back (Fig. 2-27F) to fit between the wall pieces, using the marking out under the top to obtain the width and bevels.

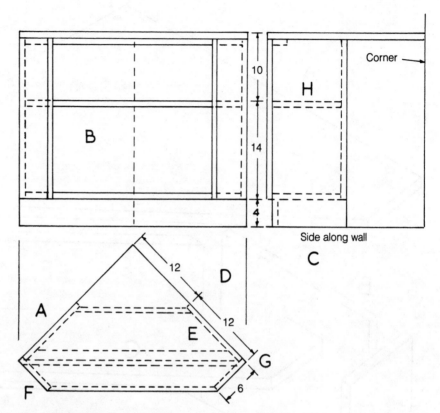

Fig. 2-26. Sizes of the corner table.

7. One shelf is suggested (Fig. 2-26H). It need not reach the front, but can be cut straight across at the angle behind the upright (Fig. 2-27G).
8. Also set the toe board back from the front and join it to the edges of the wall boards (Fig. 2-27H).
9. Most of the joints are hidden and can be screwed or nailed. You can secure the top by screwing upwards through the subtop, adding a few dowels into the back. You can screw downwards through the bottom into the toe board or use dowels.
10. It will be best to dowel the uprights to other parts where they cross to avoid screwheads showing.
11. Try the assembly in the room corner, preferably before finally fixing down the top. If the fit is satisfactory, finish the wood by staining and polishing or in any way you prefer.
12. The table should be heavy enough to stay in position unaided, but you could drive a screw in each side through the wallboards if you want to hold it in place.

44 Tables and Stands

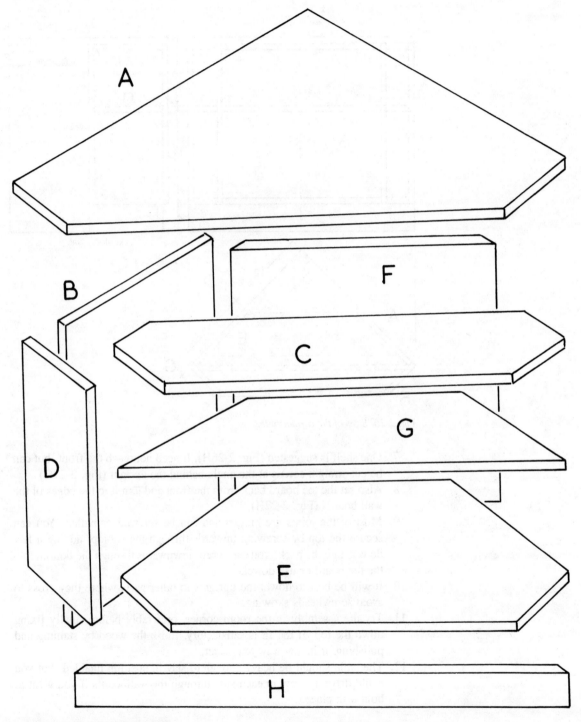

Fig. 2-27. Parts of the corner table.

Bedside Extending Table 45

Materials List for Corner Table	
1 top	24 × 24 × 3/4 plywood
2 sides	6 × 24 × 3/4 plywood
2 wall pieces	12 × 28 × 3/4 plywood
1 bottom	12 × 34 × 3/4 plywood
1 subtop	6 × 34 × 3/4 plywood
1 back	18 × 24 × 3/4 plywood
1 shelf	9 × 34 × 3/4 plywood
1 toe board	4 × 34 × 3/4 plywood

Bedside Extending Table

The table/cabinet in Fig. 2-28 is intended to stand beside a bed in the usual way, but it has a flap that can extend partly over the bed when you want to support a book, mug, or even a complete meal. The table is on casters, so it can be easily pulled into position or pushed back out of the way.

Most parts are 3/4-inch plywood. The top and flap could be surfaced with wood or plastic veneer. Edges might be covered with iron-on strips or solid wood. The suggested sizes in Fig. 2-29A should suit most beds, but you can easily alter heights; 2 inches have been allowed for the casters. It is advisable to buy and measure your pieces before you start construction. You could screw most parts if the heads will be sunk, covered, then painted over, otherwise use 5/16-inch dowels at about 4-inch centers. Most parts are simple rectangular panels. Be careful to cut squarely.

1. Start by marking out a side with the positions of other parts on it (Fig. 2-29B). The back fits inside the sides. Two shelves reach from the back to the front edge, and there is a stiffener under the front of the bottom one. The other shelf is cut back to improve the view of things on the bottom shelf.
2. Make the three shelves (Fig. 2-30A) and the back (Fig. 2-30B), which is the same height as the sides. The stiffener (Fig. 2-30C) fills the space under its shelf.
3. Mark all these parts for dowels, then glue them together while the assembly is standing on a level surface.
4. Decide which side you want the flap. Make the top (Figs. 2-29C and 2-30D) level with the sides at the back and with a 2-inch overhang at the flap side and a 1-inch overhang at the other side and front.
5. The flap (Figs. 2-29D and 2-30E) follows on the width of the top and hangs down. Three 1 1/2-inch hinges underneath should be satisfactory, but delay fitting them until you add the bracket, so they can be located where they will not affect the bracket's movement.
6. Attach the top to the other parts.

46 Tables and Stands

Fig. 2-28. This bedside table/cabinet has a flap that extends over the bed.

7. The bracket folds under the hanging flap (Fig. 2-29E) but opens to support it (Fig. 2-29F). Cut the bracket to shape (Fig. 2-29G). The cutout opening is optional, but it provides a finger grip when moving the bracket. Round all exposed edges.
8. Use two hinges on the bracket. Position the bracket towards the front of the table for easy access and so it folds back neatly. Attach the flap with its hinges.
9. You will have to provide bases at the bottom corners for the casters. What you make depends on how they have to be attached. Blocks inside the corners (Fig. 2-30F) should provide a suitable area for most casters.
10. If you use a painted finish, a light color inside will make the contents easier to see.

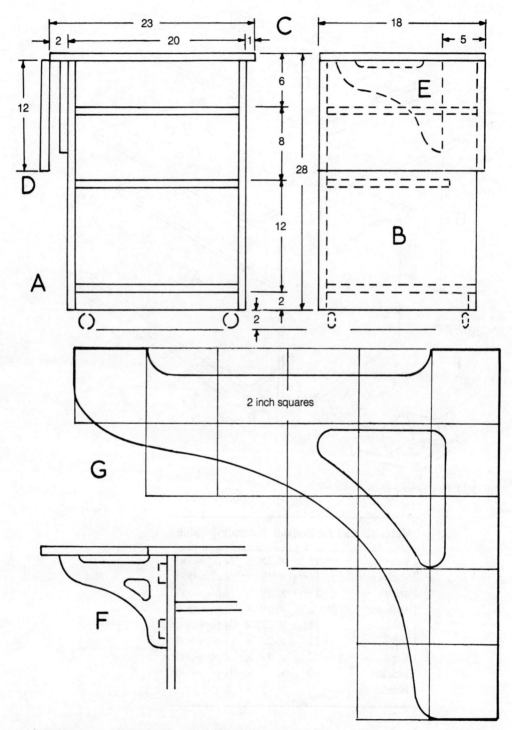

Fig. 2-29. Sizes of the bedside table and details of the bracket.

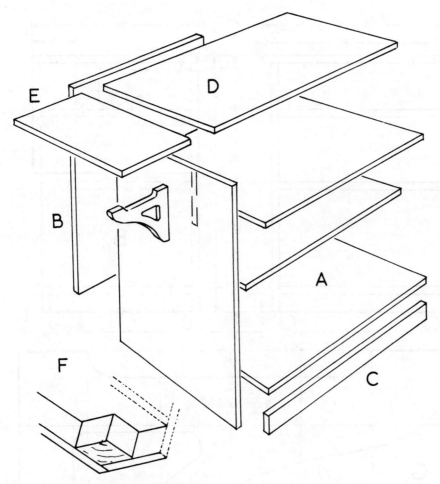

Fig. 2-30. Parts of the bedside table/cabinet.

Materials List for Bedside Extending Table		
2 sides	17	× 28 × 3/4 plywood
2 shelves	16 1/4	× 19 × 3/4 plywood
1 shelf	13 1/4	× 19 × 3/4 plywood
1 stiffener	2	× 19 × 3/4 plywood
1 back	18 1/2	× 28 × 3/4 plywood
1 top	18	× 23 × 3/4 plywood
1 flap	12	× 18 × 3/4 plywood
1 bracket	10	× 12 × 3/4 plywood
4 blocks	2	× 2 × 2

3

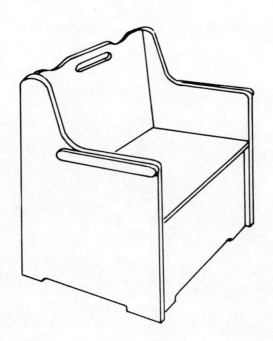

Seats and Stools

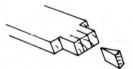

Foot Stool

The small stool in Fig. 3-1 can be used as a foot rest while sitting, as a stand to reach higher, or as a child's seat. It is decorative and reminiscent of country furniture design, although obviously, the early makers would not have used plywood.

The sizes (Fig. 3-2A) could be varied. If you alter them, make sure the bottoms of the legs extend almost to the ends of the top, for stability. For the sizes suggested, you could use any thickness from $1/2$-inch to $3/4$-inch plywood. Much depends on the plywood. A hardwood plywood with many veneers in the thick-

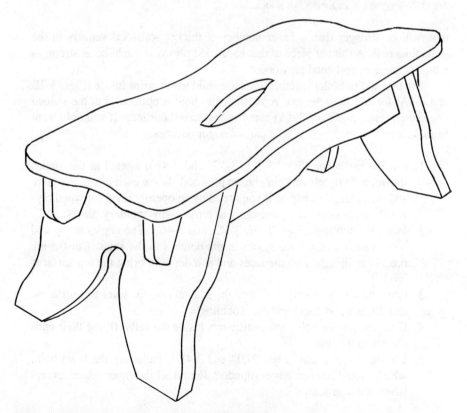

Fig. 3-1. This foot stool has shaped edges and a hand hole for lifting.

52 Seats and Stools

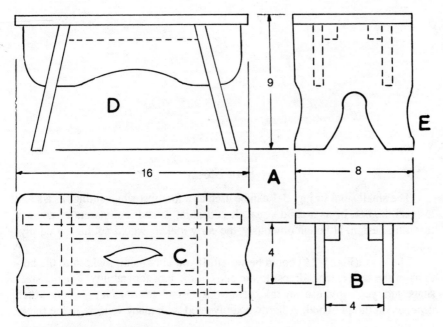

Fig. 3-2. Suggested sizes for the foot stool.

ness will be stronger than a fewer number of thicker softwood veneers in the same thickness. A thinner piece of that hardwood plywood might be as strong as a thicker piece of softwood plywood.

The legs and rails slot together and have solid wood strips inside (Figs. 3-2B and 3-3A) for attaching the top. A central hand hole is optional, but the shaped one shown (Figs. 3-2C and 3-4A) matches the curved outlines. If you only want to make a utility stool, give all the parts straight outlines.

1. Cut the two sides (Figs. 3-2D, 3-3B, and 3-4B), slotted at the angles shown, to fit tightly on your chosen plywood. Leave each top edge square and round all other edges. If you are using an open-grained softwood plywood, sand edges very thoroughly to remove any splintery pieces.
2. Make the two legs (Figs. 3-2E, 3-3C, and 3-4C). The angles at top and bottom should match the slopes of the notches in the sides. Cut the leg notches to fit tightly on the sides and at a depth to bring the top surfaces level.
3. Glue these parts together. Check for squareness, for level top surfaces, and for feet that stand without wobbling.
4. Glue and pin the solid wood stiffeners inside the rails, fitting their ends closely to the legs.
5. Cut the top to shape (Figs. 3-3D and 3-4D), including the hand hole, which should have its edges rounded. Round all the outer edges, except where the legs join.

Foot Stool

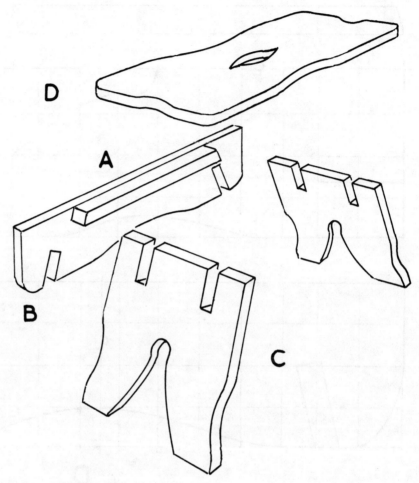

Fig. 3-3. Parts of the foot stool.

6. Glue and screw the top downwards into the solid wood strips and add a few pins into the legs and rails. If you do not want screwheads to show on top and if the plywood is thick enough, drive screws upwards. In any case, if the parts fit well, glue will contribute considerable strength.
7. Finish the stool with varnish or paint. Having the top a different color from the other parts produces an interesting effect.

Materials List for Foot Stool	
2 rails	4 × 16 × 5/8 plywood
2 legs	8 × 10 × 5/8 plywood
1 top	8 × 17 × 5/8 plywood
2 stiffeners	7/8 × 7/8 × 12

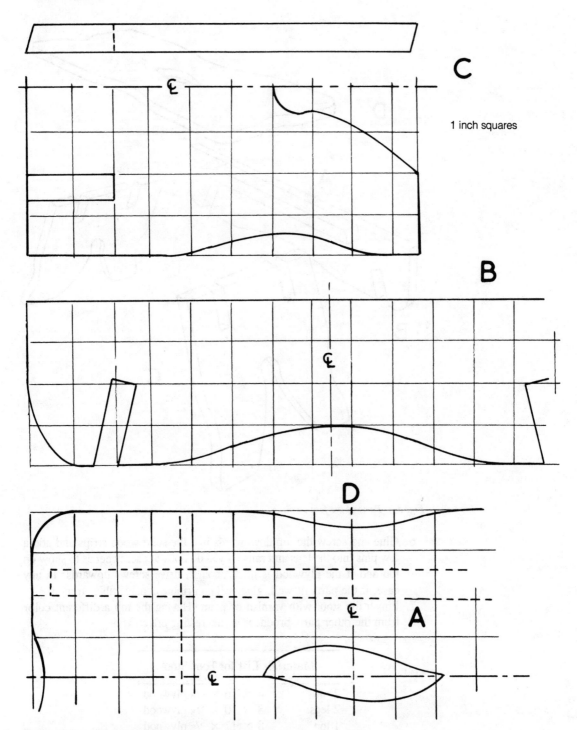

Fig. 3-4. The shaped parts of the foot stool.

Child's Chair

A toddler can use a chair scaled down to his or her size that will stand up to rough treatment. The chair might be treated as a toy as much as a piece of furniture. The chair in Fig. 3-5 is intended to serve as a strong play seat, not as a permanent piece of living room furniture. It has arms to prevent the child falling sideways, and the back, which is high enough to provide support, has a hand hole for lifting. Underneath the seat is a compartment with a door at the back, which should add interest for the young user.

Construction is with 1/2-inch plywood and solid wood square strips, all glued and pinned. There is no need to cut joints between parts of the framing, all of which is hidden. Paint in a bright color would be the best finish, and you could

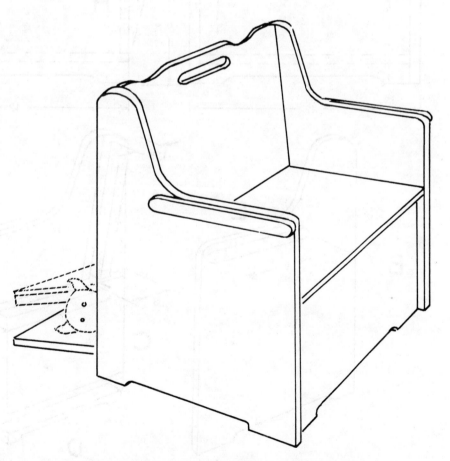

Fig. 3-5. The child's chair has storage under the seat and a door at the back.

56 *Seats and Stools*

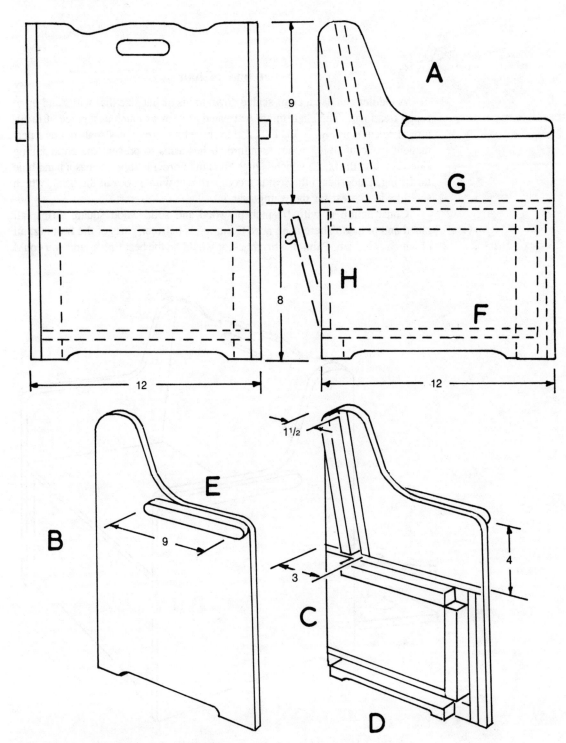

Fig. 3-6. Sizes and framing of the child's chair.

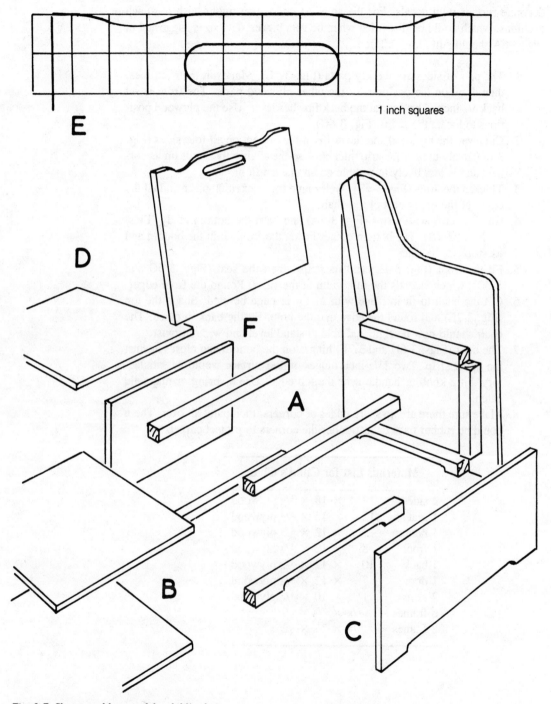

Fig. 3-7. Shapes and layout of the child's chair.

decorate further with decals. For the smallest child, you might wish to attach cushions, which could be discarded when he gets bigger. The sizes suggested in Fig. 3-6A should suit most children.

1. The pair of sides are the key parts (Fig. 3-6B). Mark out their outlines, then draw on the positions of the other plywood parts. The front is set back 1/2 inch. The door at the back finishes level. Use the plywood positions to locate the strips (Fig. 3-6C).
2. Cut away the plywood and the strips at the bottom on all four sides (Fig. 3-6D) to about half the strip thickness, so the chair only stands on its corners and is less likely to wobble on an uneven floor.
3. Thicken the arms (Fig. 3-6E). Well round the thickening piece and all the edges of the upper part of the chair.
4. Cut four crosswise strips (Fig. 3-7A) and have the bottom ready (Figs. 3-6F and 3-7B). The bottom comes inside the front. Join the bottom and its strips to the sides.
5. Fit the front (Fig. 3-7C) and its strips, then the seat (Figs. 3-6G and 3-7D), which extends the full width of the sides. Round the front edge.
6. Cut the back to fit in place with its lower edge beveled. Shape the top (Fig. 3-7E) and round the edge and the hole. Fit the back in place. The chair should now be rigid and able to stand level and without twist.
7. The door (Figs. 3-6H and 3-7F) hinges on the bottom and closes against the upper strip. Two 1 1/2-inch hinges on the surface would be satisfactory. Fit a knob or handle near the top edge. Use a spring or magnetic catch.
8. Make sure there are no sharp edges or corners, then paint all over. There could be rubber pads or feet under the corners to protect carpets.

Materials List for Child's Chair

2 sides	12	× 18 × 1/2 plywood
1 seat	12	× 12 × 1/2 plywood
1 bottom	12	× 12 × 1/2 plywood
1 front	8	× 12 × 1/2 plywood
1 back	10	× 12 × 1/2 plywood
1 door	7	× 12 × 1/2 plywood
2 arms	1 1/2	× 10 × 1/2 plywood
6 frames	7/8 × 7/8 × 9	
8 frames	7/8 × 7/8 × 12	

Armchair

The armchair in Fig. 3-8 could be made from exterior plywood for use outdoors, or it could be finished in hardwood plywood for use in a living room. Any plywood with a painted finish would be suitable for a den or playroom. In all cases, the chair could be used as it is or with loose cushions.

All main parts are 3/4-inch plywood with 1-inch square wood strips strengthening the joints. The feet are thickened (Fig. 3-9A) to reduce sinking in soft ground or marking carpets. Key measurements are 24 inches. It should be possible to cut the plywood parts (Fig. 3-9B) from a standard sheet.

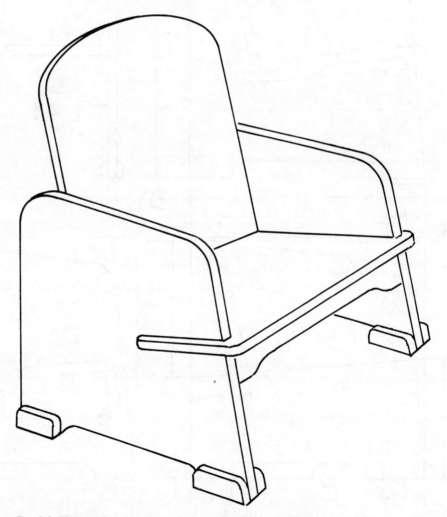

Fig. 3-8. This armchair has plywood parts with solid wood stiffening.

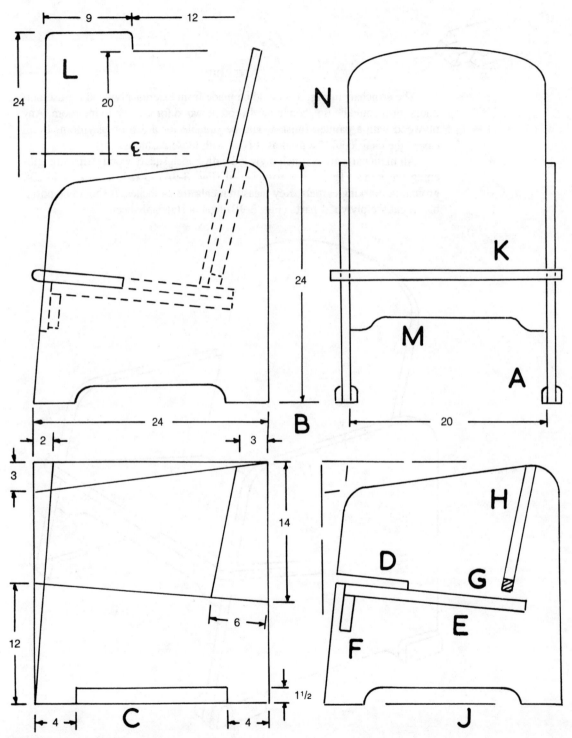

Fig. 3-9. Sizes and how to set out the armchair.

1. The chair sides are based on 24-inch squares. Mark the pair of sides with the lines shown (Fig. 3-9C). These give the outline and the location of other parts.
2. On each piece at the seat line, cut a notch 8 inches deep and a close fit on the thickness of the plywood (Fig. 3-9D). Fix a strip below this line extending 12 inches beyond the notch and stopping 3/4 inch from the front edge (Fig. 3-9E). Below this strip put a 3-inch strip to fit behind the seat support (Fig. 3-9F).
3. The back will meet the seat and have a strip at the joint (Fig. 3-9G). Allow for the thickness of the seat plywood and this strip, then fit a joint strip (Fig. 3-9H).
4. Shape the bottom cutout (Fig. 3-9J). If the feet are to be thickened, put blocks of plywood or solid wood on each side. Round their outer edges and corners. Round the top corners of the sides.
5. The seat (Fig. 3-9K) has a front to fit the side notches and extend 1 inch forward (Fig. 3-9L). Round the outer corners and all exposed edges.
6. Make the front seat support to fit between the sides (Fig. 3-9M). Round the edges of its shaped underside.
7. The back is 24 inches high and the same width as the reduced part of the seat (Fig. 3-9N). Shape its top in any way you wish.
8. Bevel a strip in readiness for going across the bottom of the back when you assemble.
9. Fit the seat between the sides first, with glue and screws into the solid wood strips. Fit the front seat support close under the seat, screwing through it and into the short strips. Add the back between the sides with similar screwing. This should draw the chair square, but check that it is free from twist by testing it on a flat surface.
10. Well round the front edges and corners of the arms and take sharpness off all other edges and corners before finishing with paint or varnish.

Materials List for Armchair	
2 sides	24 × 24 × 3/4 plywood
1 seat	21 × 24 × 3/4 plywood
1 back	21 × 21 × 3/4 plywood
1 seat support	5 × 21 × 3/4 plywood
5 strips	1 × 1 × 21

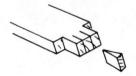

Side Chair

A simple side chair is always useful. You can never have enough of them. You might need one for extra guests at a dining table, or one to use elsewhere when doing laundry, playing a game, or working at a hobby. A chair that is not as valuable as your best dining chairs can be pressed into use for a great many purposes where rougher treatment is a possibility.

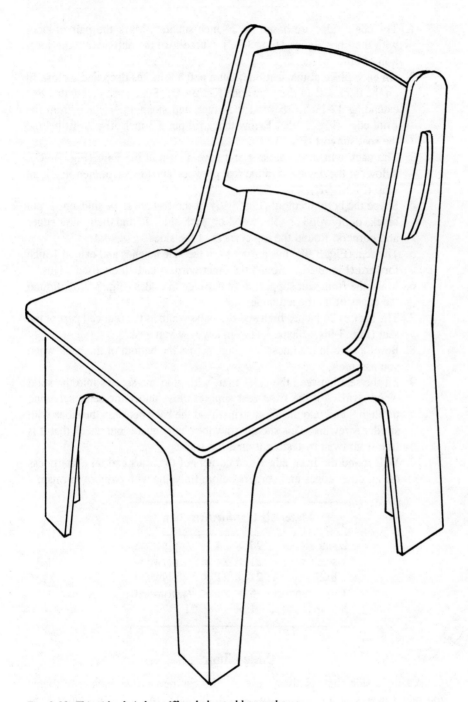

Fig. 3-10. This side chair has stiffened plywood legs and seat.

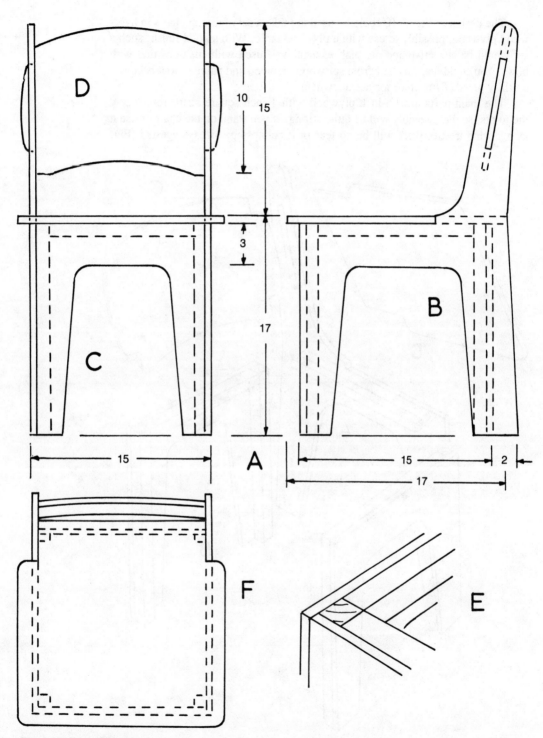

Fig. 3-11. Sizes of the side chair.

64   *Seats and Stools*

The design in Fig. 3-10 could be used to make a set of dining chairs in hardwood plywood, possibly to use with a plywood table. With a good finish, such a set would be attractive and unusual, especially if used with cushions tied with bows. You could make a chair from softwood plywood and paint it, so it becomes a utility piece of furniture for use anywhere.

The main parts are 1/2-inch plywood with 1-inch square strips reinforcing the joints, so the assembly will be quite strong. If you stand on the chair or use it as a sawing trestle, there will be no fear of it collapsing. Sizes suggested (Fig.

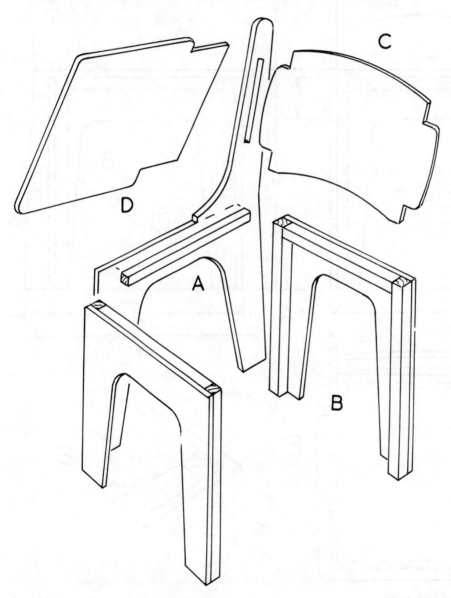

Fig. 3-12. The chair parts, showing its framing.

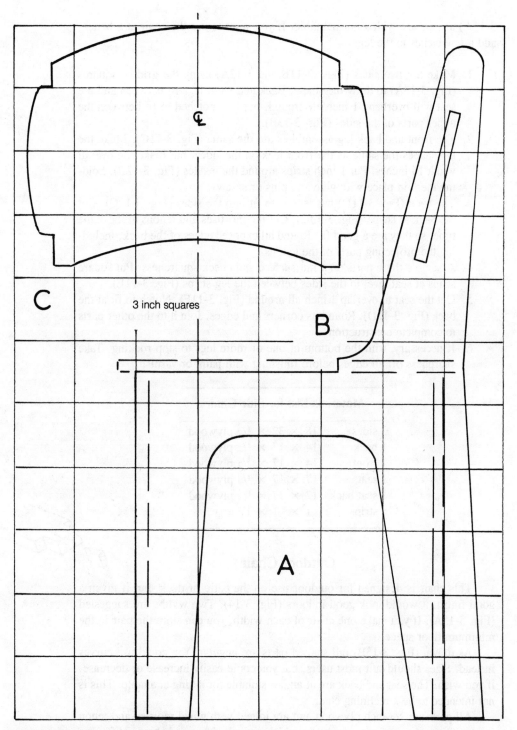

Fig. 3-13. Shapes of the back and sides of the side chair.

3-11A) are for use with a normal table. If you want to use the chair at a bench, add a few inches to the legs.

1. Make the two sides (Figs. 3-11B and 3-12A) using the grid of squares (Fig. 3-13A) to draw the shape. Back and front legs fit between and the seat will overhang 1 inch all around, but it is notched to fit between the upper parts of the sides (Fig. 3-13B).
2. The front and back leg assemblies are the same (Fig. 3-11C). Make the leg shapes the same as the front legs on the sides, but make the overall width 14 inches. Put 1-inch strips around the insides (Fig. 3-12B), holding them in place with glue and pins or screws.
3. The back (Fig. 3-11D) has tenons to fit into the sides (Fig. 3-12C). Cut the back to shape (Fig. 3-13C). Cut the mortises and tenons at the same time, so they are a good fit. Round all exposed edges of the back, including the projecting parts of the tenons.
4. Assemble these parts on a flat surface and check squareness. Put square strips at seat level in the sides between the leg strips (Fig. 3-11E).
5. Cut the seat to overlap 1 inch all around (Fig. 3-11F). Notch to fit at the back (Fig. 3-12D). Round its corners and edges. Join it to the other parts to complete construction.
6. If necessary, trim the bottom of one or more legs to stop rocking. Take sharpness off all edges before finishing with paint or varnish.

Materials List for Side Chair	
2 sides	18 × 32 × 1/2 plywood
1 back	14 × 17 × 1/2 plywood
1 front	14 × 17 × 1/2 plywood
1 seat	17 × 17 × 1/2 plywood
1 seat back	12 × 17 × 1/2 plywood
6 strips	1 × 1 × 17

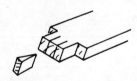

Outdoor Chair

This chair is designed for outdoor use on the patio or deck, but if given a good finish, it would look good indoors (Fig. 3-14). Two widths are suggested (Fig. 3-15A). If you make one chair of each width, you can store the pair in the minimum floor space.

As drawn (Fig. 3-15B), all shaped parts are angular. You could use curves instead. Sizes should suit most users, but you could easily increase or decrease, if you wish. The seat and back are at angles suitable for sitting at a table. This is not intended to be a reclining chair.

All parts are 3/4-inch plywood and are joined with glued mortise-and-tenon joints. The seat tenons into the back, which extends down and shares stiffening

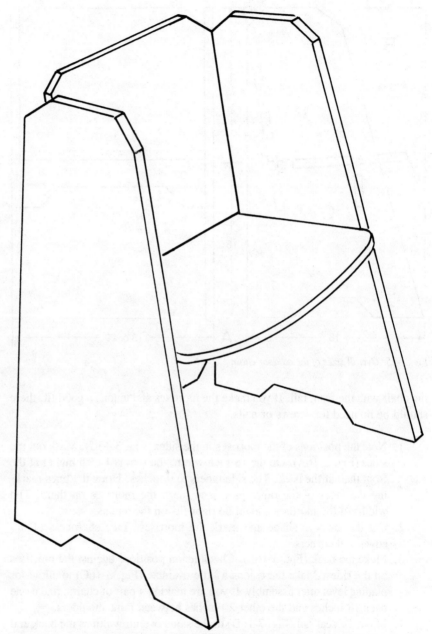

Fig. 3-14. This substantial chair is intended for use on a patio or deck.

68 Seats and Stools

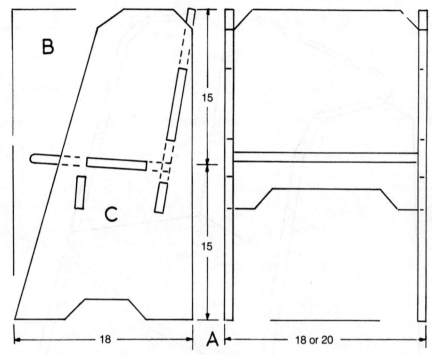

Fig. 3-15. Overall sizes of the outdoor chair.

the chair with the front rail. If you make the mortises and tenons a good fit, there should be no need for screws or nails.

1. Note the positions of the mortises in the sides (Fig. 3-15C). Mark out the sides (Fig. 3-16A). On the 18-inch width, the seat is 1 inch higher at the front than at the back. The side tapers to 9 inches. Draw the lines marking the faces of the other parts and mark the mortises on them. The widths of the mortises should be tight fits on the tenons.
2. Cut the sides to shape and mark the mortises. Take sharpness off all edges and corners.
3. Make the back (Fig. 3-16B). Check tenon positions against the mortises on the sides. Make the tenons a little oversize (Fig. 3-16C) to allow for planing level after assembly. If you are making a pair of chairs, make one back 18 inches and the other 20 inches between joint shoulders.
4. Make the seat rail (Fig. 3-16D) to the same overall width as the back and with matching tenons and cutout.
5. Take sharpness off what will be exposed edges on these parts.
6. Make the seat (Fig. 3-16E) to a matching width. Check the edge tenons against the mortises in the chair sides. The rear tenon has to match the mortise in the back. They do not cross quite squarely, but the difference is slight and you might have to ease the mortise during assembly. Well round the front edge of the seat.

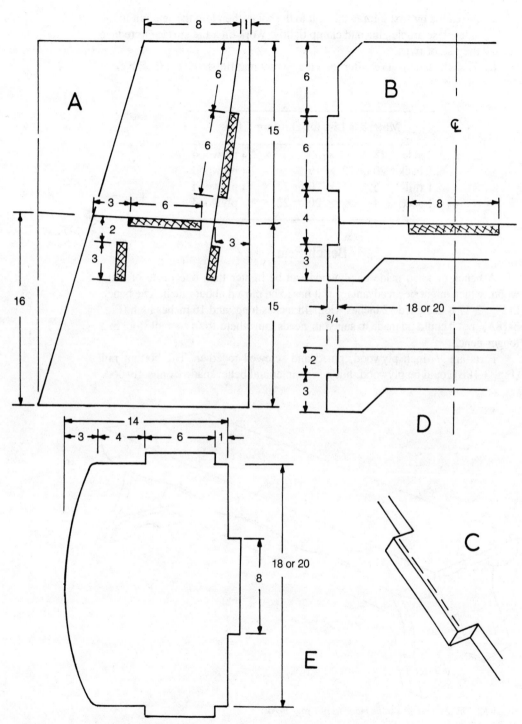

Fig. 3-16. Sizes of parts of the outdoor chair.

7. Assemble by first joining the seat to the back, then join the seat rail to the sides. Use ample glue and clamp tightly. Work on a flat surface to reduce the risk of twist.
8. Trim the tenons level after the glue has set and finish the chair with exterior grade paint.

Materials List for Outdoor Chair

2 sides	18	× 30	× 3/4 plywood
1 back	20 or 22	× 19	× 3/4 plywood
1 rail	5	× 20 or 22	× 3/4 plywood
1 seat	14	× 20 or 22	× 3/4 plywood

Bench Seat

A bench or form made of plywood can be lighter than one made of solid wood, which might be an advantage if it has to be moved about much. The bench in Fig. 3-17 is designed 12 inches wide, 48 inches long, and 16 inches high (Fig. 3-18A), but it could be made to suit your needs, anywhere from a small stool to a longer bench.

Parts are 3/4-inch plywood, glued and screwed together. The bottom rail (Fig. 3-18B) could be plywood, but it is simpler and better made from softwood.

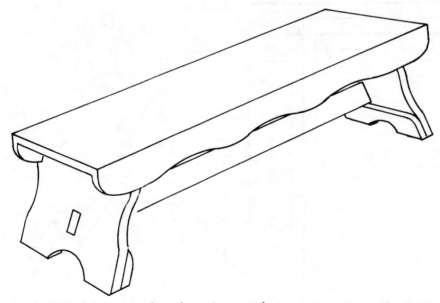

Fig. 3-17. This bench seat can be made to suit your needs.

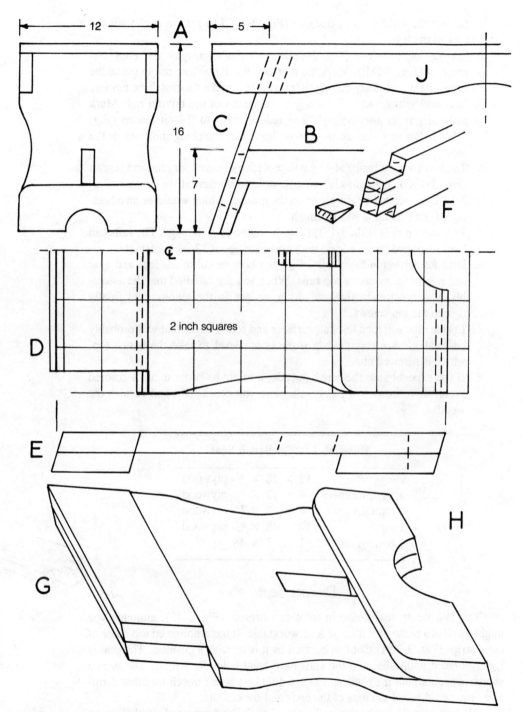

Fig. 3-18. Suggested sizes for the bench seat.

72 Seats and Stools

1. Set out the main lines of one end (Fig. 3-18C) to get the angles for the cuts in the legs.
2. Set out the two legs (Fig. 3-18D) with the angles for the ends and recesses (Fig. 3-18E). Mark the mortises for the bottom rail to cut to the same angle, but delay cutting them until you make the matching tenons.
3. Your end setting out will show you the length of the bottom rail. Mark and cut out its angular ends and reduce them to 2-inch tenons (Fig. 3-18F). Put saw cuts across for wedges. Cut matching mortises in the legs.
4. Thicken each leg inside at its top to provide extra area for glue and screws (Fig. 3-18G). It might also be advisable to thicken at the bottom (Fig. 3-18H) to increase bearing area on the ground, which would be an advantage if the ground is soft or rough.
5. The two top rails (Fig. 3-18J) fit in the recesses in the legs. The rails can have scalloped lower edges, with curves at about 12-inch intervals.
6. Drill for screws in these rails. Glue the bottom rail to the legs and glue and lightly screw on the top rails. When you are satisfied that the assembly is symmetrical, glue and drive wedges in the tenons, and finally tighten the top screws.
7. Let the glue set, then level all surfaces and round edges that will probably be handled. See that the top surfaces are level to take the bench top, which is screwed on.
8. If you expect to use the bench outside, it might be better to use a colored preservative instead of paint, otherwise you could stain and varnish the wood.

Materials List for Bench Seat

2 legs	12 × 18 × 3/4 plywood
4 leg thickeners	4 × 12 × 3/4 plywood
2 top rails	4 × 48 × 3/4 plywood
1 top	12 × 48 × 3/4 plywood
1 bottom rail	1 × 3 × 48

Dresser Seat

This is a vanity seat to use in front of a dresser (Fig. 3-19), although you might use it at a piano or with a desk or worktable. It has a hollowed top made of three strips (Fig. 3-20A) that can be used as it is or with a cushion. The seat is rigidly braced with rails, and the ends have hand holes for lifting. The design shown is angular, but if a rounded shape would be a better match for other furniture, you could round the tops of the ends and the cutouts.

All parts are 3/4-inch plywood, which could be hardwood, with iron-on strips on the edges, then a clear finish. You could also use softwood plywood and paint it. It is the pair of ends that controls the other sizes, so make them first.

Dresser Seat 73

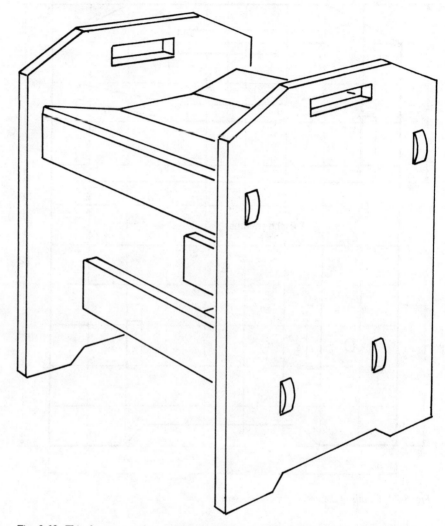

Fig. 3-19. This dresser seat has a shaped top that can be used with or without a cushion.

1. Mark out the ends (Figs. 3-20 and 3-21A). Mark on the shape and position of the seat. Two rails (Figs. 3-20B and 3-21B) fit each side of the seat support (Figs. 3-20C and 3-21C). Two similar rails brace the sides lower down (Figs. 3-20D and 3-21D).
2. Shape the outlines. Cut the hand holes and round the edges. Mark the mortises.
3. Make four rails 16 inches between the shoulders (Fig. 3-21E), with tenons that will project through the sides a little with rounded ends (Fig. 3-21F). Cut the mortises in the ends to match the tenons.
4. Cut the seat supports to fit between the top rails, with 1-inch rise at the sides. Glue and screw them to the ends.

74 Seats and Stools

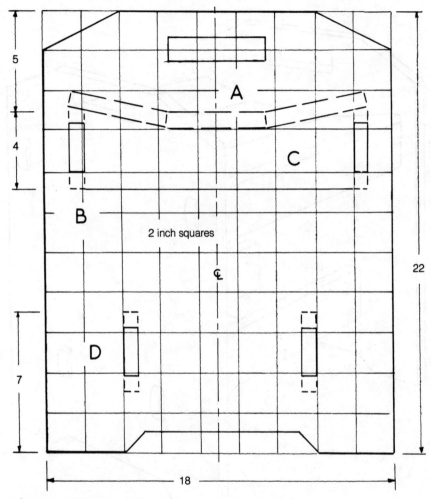

Fig. 3-20. Details of the end of the dresser seat.

5. Join the ends with the four rails. Check for squareness and lack of twist. Clamp tightly.
6. Cut the three seat strips (Fig. 3-21G). Miter them to fit closely or to leave narrow gaps, if you wish. Round the exposed edges of the outer ones. Glue and screw the strips to their supports.
7. Well round exposed edges before applying a finish.

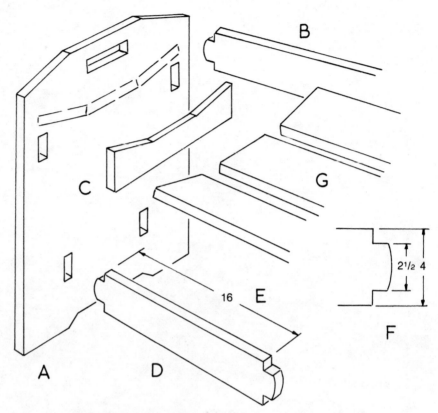

Fig. 3-21. Assembly of the dresser seat.

Materials List for Dresser Seat	
2 ends	18 × 22 × ³/₄ plywood
4 rails	4 × 19 × ³/₄ plywood
2 seat supports	4 × 15 × ³/₄ plywood
3 seat strips	5 × 17 × ³/₄ plywood

4

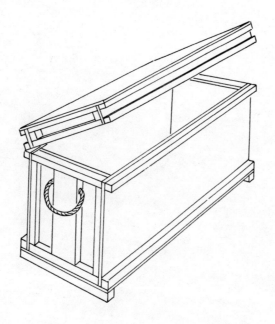

Containers

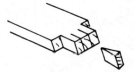

Chest

The chest in Fig. 4-1 is a substantial storage box. It is not intended to be a piece of furniture. It is more suited for tools in the workshop or to take to a site, for outdoor games equipment, or for storing things together in the loft or basement. It will keep the contents tidy and protect them from dirt and damp.

The chest is shown with all its framing outside (Figs. 4-2 and 4-3). This gives a smooth interior, which will not snag anything you put in. This might be important if you are storing fabrics. It also facilitates the fitting of compartments or trays for tools, hardware, or other small items. If you prefer a smooth exterior, the box can be made inside-out, with the end parts turned inwards and the sides framed inside.

The chest can be made with any plywood from $1/4$ inch upwards, but it is best made of $1/2$-inch plywood for general purposes. If you use external or marine plywood, the chest will not suffer if left outdoors. All the framing is made of $3/4$-inch square strips, which could be hardwood or softwood. A hardwood plywood with matching hardwood strips would look good if varnished. For a painted finish, you could use any wood, and it would not matter if they are mixed.

There is no need to cut joints at the corners of framing parts. If you use a good glue and pins or fine nails closely spaced (Fig. 4-3A), the framing strips will be held by the plywood. At the corners of the meeting opening parts, extra strength is provided by dowels (Fig. 4-3B).

Sizes are suggested (Fig. 4-3C), but you can use the same constructional method for chests that are bigger or smaller. The box and lid are first made as one unit, then separated after assembly. This ensures an exact match.

1. The key parts are the pair of ends (Figs. 4-2A and 4-3D). Cut matching plywood pieces and attach the square strips. Leave a gap for sawing and planing after separating the lid from the box (Fig. 4-3E). Check the width of cut made by the saw you will use and allow for a few strokes with a plane—probably a gap of $3/16$ inch will do.
2. Glue and pin these strips in place. At the strips each side of where the cut will come, keep the pins clear of where you will drill for dowels during assembly.
3. The wide piece at the center of each end is to take a rope strop handle. Rope about a $3/8$-inch diameter will suit. Drill the wood for the rope (Fig. 4-3F) before fitting the piece into the frame.

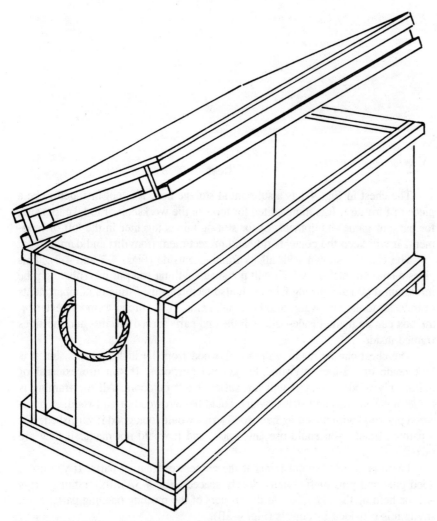

Fig. 4-1. This chest is made of plywood that is framed outside to give a smooth interior.

4. Make the two sides (Fig. 4-2B) long enough to overlap the ends. Put strips on them to match the spacing of the horizontal strips on the ends.
5. Glue and pin or screw the sides to the ends. Use 1/4-inch dowels (Fig. 4-3B) at the corners of the close strips. You could use more dowels at top and bottom, but the plywood there should provide enough strength.
6. Add the top (Fig. 4-2C) and bottom (Fig. 4-2D). Strips across the ends of the bottom (Fig. 4-2E) will raise the chest off the floor and keep it clear of damp ground. True up and sand all exterior surfaces.
7. Saw between the strips to separate the lid from the box. Mark the pieces so that they will not be reversed during final assembly. Plane the meeting edges.
8. Let in hinges along the rear edges. For the sizes suggested, two 3-inch

Chest 81

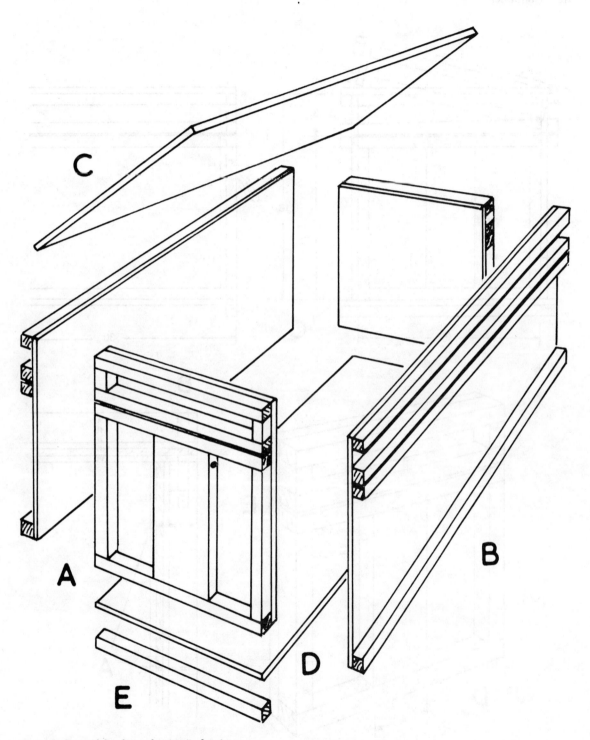

Fig. 4-2. Parts of the chest, showing its framing.

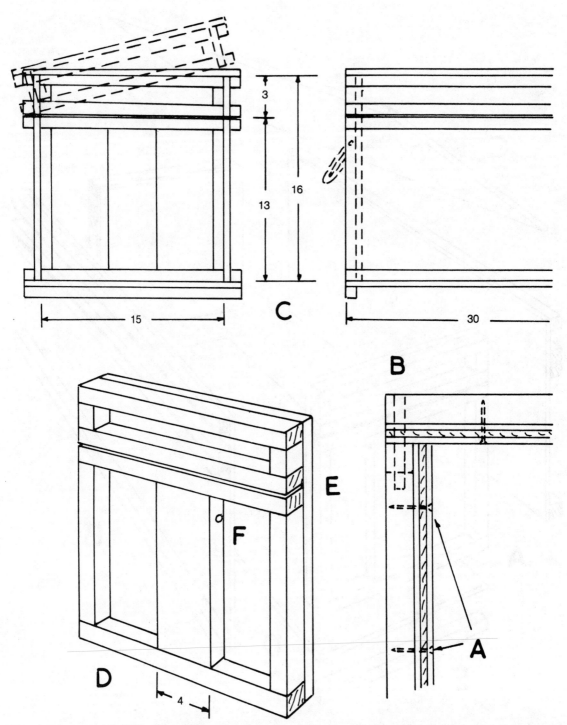

Fig. 4-3. Sizes and construction of the chest.

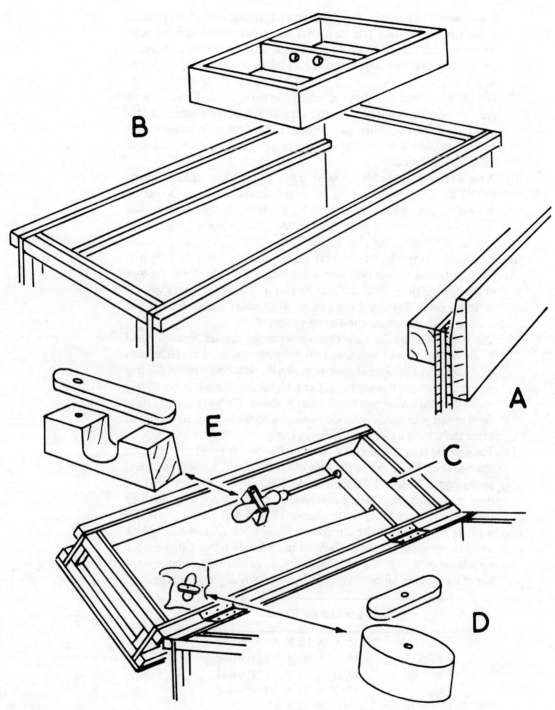

Fig. 4-4. Fitting out the chest with tray and toolholders.

hinges would be suitable. You could have a hasp and staple for a padlock at the front or let in a box lock. You might have to thicken the edge where the lock comes. If you do not want to let the lid swing right back, fit a folding strut or a rope between eyes inside one end. Make the rope loop handles.

9. Step #8 completes the construction of the basic chest. Finish it with paint or varnish. Paint outside and varnish inside looks good, or you could use a lighter-colored paint inside. Even if the main finish is varnish, paint underneath will give better protection if the chest will have much use outdoors.
10. A plain chest might be all you need, but it is possible to make compartments, trays, or attachments inside to suit special contents. Do not be too enthusiastic with what you do inside, however, and be sure you really need it, because parts fitted inside take up space. If you want maximum capacity, leave the inside plain.
11. You could arrange the lid to lift off instead of swing open. To keep it in place, put strips around the inside of the box (Fig. 4-4A). Taper the tops of the strips slightly, so the lid goes over easily but settles tightly when it is fully closed. You could use similar strips under a hinged lid if you want to make the closed chest more dustproof.
12. One useful addition for many chests is a tray for smaller items. It could be the full length and width, and lift out when you need to reach items below it, or it could be shorter (Fig. 4-4B). You could make the tray edge level with the box top or project it higher into the lid. A boxed tray with a division having two finger holes is shown, but you can make divisions to suit your needs. Provide two strips on the sides to act as slides. Make the tray an easy fit, so it moves freely.
13. The deep lid is a good place for storing tools. Slot and drill a strip across one end (Fig. 4-4C) to take the ends of saws, chisels, screwdrivers, and similar tools. To secure the tools, make blocks to take handles and lock them with turnbuttons. For a saw handle, make a block to go through the hand hole and fit a thin turnbutton (Fig. 4-4D) with a screw and washer. For round handles, make a hollow block (Fig. 4-4E), which could be extended to have hollows for many handles. Put a single-sided turnbutton on top. With a little ingenuity, you can arrange a large number of tools in the lid by sharing fittings and partially overlapping.

Materials List for Chest

2 ends	15 × 16 × 1/2 plywood
2 sides	16 × 30 × 1/2 plywood
1 bottom	17 1/2 × 30 × 1/2 plywood
1 top	17 1/2 × 30 × 1/2 plywood
12 end frames	3/4 × 3/4 × 17
8 side frames	3/4 × 3/4 × 32
2 feet	3/4 × 3/4 × 19

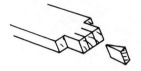

Umbrella Stand

The container in Fig. 4-5 is large enough to take all the canes, umbrellas, even fishing poles that your family might wish to put inside the front door. It is a stand that is unlikely to be accidentally moved or knocked over, but if you want to move it, the heart cutouts serve as hand grips (Fig. 4-6A).

Construction is novel. The four sides are identical, and their tenons provide decoration as they project in alternate directions. The main parts are $1/2$-inch plywood. The bottom is hidden inside, and there are four feet to provide stability and floor clearance.

1. Mark out one side (Fig. 4-6B). The mortises and spaces between them are $4^{1}/_{2}$ inches deep. The tenons extend $1/2$ inch through the mortises and have rounded ends. The other edge is shaped to a maximum of 1 inch outside each mortise. A heart cutout about 3 inches across allows several fingers to be put in for lifting. You might wish to use another design, such as an initial or something different on each piece.
2. Cut four identical sides. Make the mortises a close fit on the tenons. If you cut the mortises with a router, the tenon edges could be rounded to match.
3. Try the parts together (Fig. 4-7A). You might have to mark joints if there are slight differences in the fit between corners.
4. Cut the bottom to fit inside (Figs. 4-6C and 4-7B) and make a solid wood frame to glue below it (Fig. 4-7C).
5. Assemble the four sides and the bottom with glue.
6. Make and fit four 3-inch square feet (Figs. 4-6D and 4-7D).
7. Finish with paint or varnish. If you expect water to drip inside, use a waterproof paint inside the bottom and at least 6 inches up the sides.

Materials List for Umbrella Stand	
4 sides	14 × 30 × $1/2$ plywood
1 bottom	12 × 12 × $1/2$ plywood
4 bottoms	1 × 1 × 12
4 feet	3 × 3 × $1/2$ plywood

86 Containers

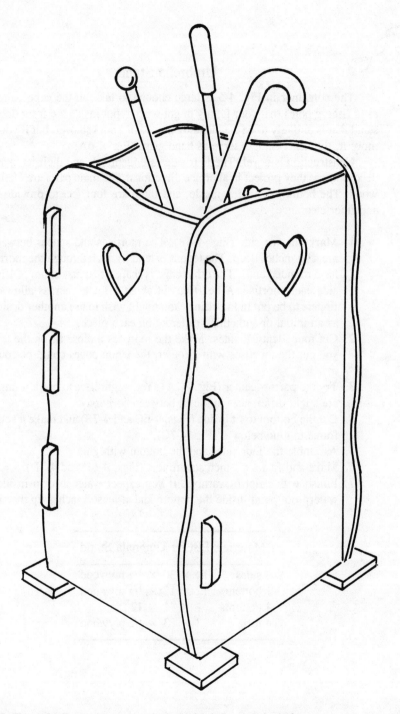

Fig. 4-5. This umbrella stand is made with plywood panels tenoned together.

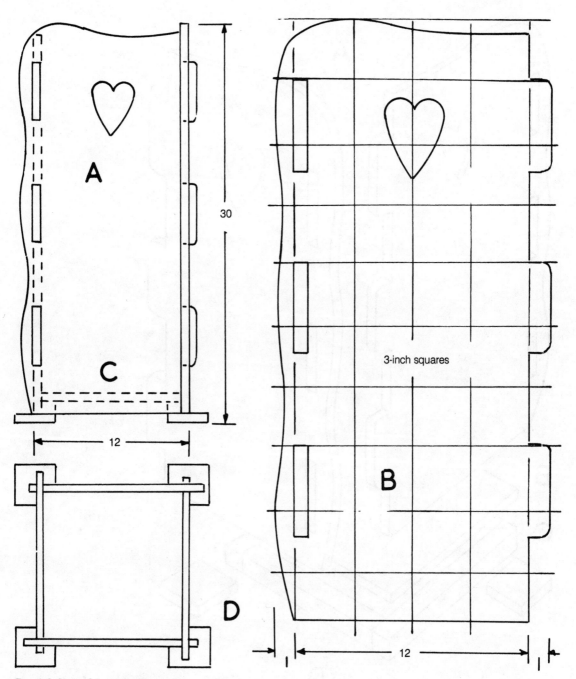

Fig. 4-6. Sizes of the umbrella stand.

88 Containers

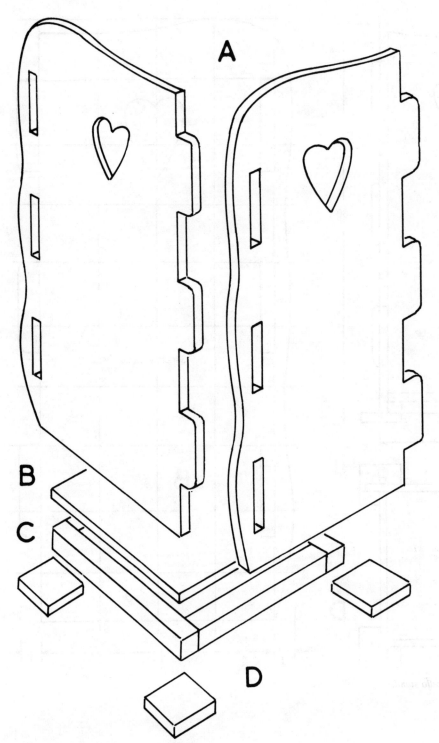

Fig. 4-7. Assembly of the umbrella stand.

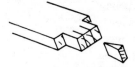

Tool Tote Box

If you want to take tools from the shop to work on some part of your property, or if you want to take small gardening tools and equipment to a further part of your yard, you need a container in which to carry them. The tote box in Fig. 4-8 has a good capacity. There is a lift-out tray for small items. The handle, which is removable, could be your gardening dibbler or just a piece of dowel rod. You could also lift the box by the framing on the ends.

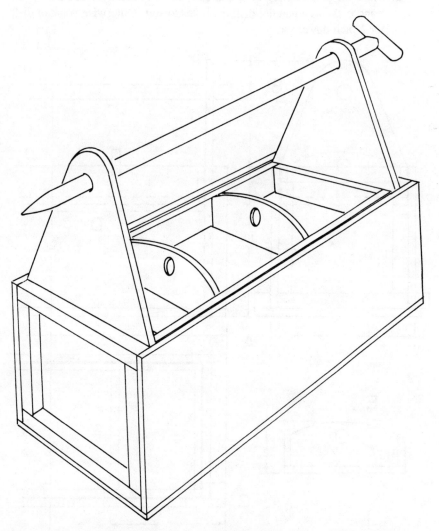

Fig. 4-8. The tool tote box can hold many types of household and garden tools in its tray and base.

Sizes are not crucial, and you might wish to make them to suit particular tools. Those suggested in Fig. 4-9A are for a box 12 inches wide, 24 inches long, and 17 inches high. The tray is 2 inches deep and has raised divisions with finger holes for lifting out. The tray will come out with the handle in place, but if you withdraw the handle, removing a particularly full tray is easier. All plywood parts can be 1/2 inch thick, and the solid wood strips are 1-inch square. All joints can be glued and nailed.

1. Mark out an end (Figs. 4-9B and 4-10A). Mark on it the outside framing (Fig. 4-10B) and the outline of the tray, as guides to further construction.
2. Make the pair of ends. If you already have a dibbler or other tool you want to use as a handle, drill the holes to suit. Otherwise you can drill for 1-inch dowel rod.

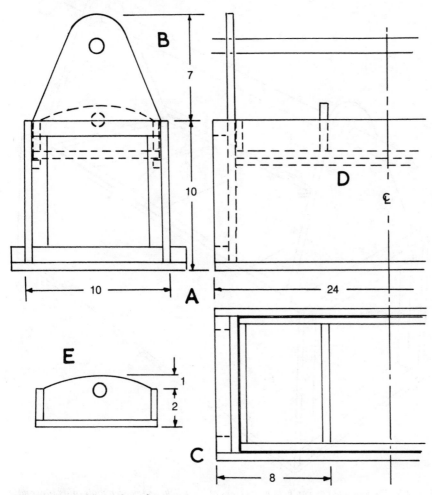

Fig. 4-9. Sizes of the tool tote box.

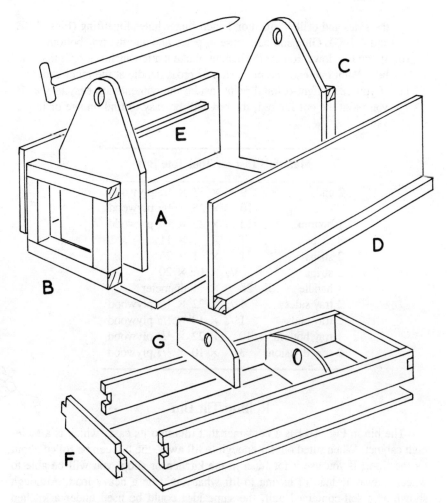

Fig. 4-10. Parts of the tool tote box and its tray.

3. Reinforce outside the ends with the solid wood strips (Fig. 4-10C).
4. Make the two sides (Fig. 4-10D) with strips along their lower edges. Join these to the ends to overlap the end strips (Fig. 4-9C).
5. Add 1/2-inch square strips to support the tray 2 inches from the top of the sides (Figs. 4-9D and 4-10E).
6. Glue and nail on the bottom. If the box is expected to be used on damp ground, put strips across under the ends to lift the bottom.
7. Make the tray to fit loosely into the box. Arrange the divisions as you wish, but the tray is easier to lift out if divisions are arranged with the center space wider than the end ones.
8. Cut finger joints at the corners (Fig. 4-10F). You can then drive fine nails both ways.
9. Make the tray divisions to fit between the sides. Curve to 1 inch above

the sides and drill 3/4-inch or 1-inch finger holes for lifting (Figs. 4-9E and 4-10G). Glue and nail these in place and add the tray bottom.

10. If you use dowel rod for the handle, make it a few inches longer than the box. Point one end and put a stop or cross handle at the other end.
11. If you use a light-colored paint inside, the contents are easy to see. If you paint the outside red, the box will be easy to find on site or in the yard.

Materials List for Tool Tote Box

2 ends	9	× 17 × 1/2 plywood
2 sides	10	× 25 × 1/2 plywood
1 bottom	12	× 25 × 1/2 plywood
8 strips	1	× 1 × 11
2 strips	1	× 1 × 25
2 strips	1/2	× 1/2 × 20
1 handle	28	× 1 diameter
2 tray sides	1 1/2	× 22 × 1/2 plywood
2 tray ends	1 1/2	× 10 × 1/2 plywood
1 tray bottom	10	× 22 × 1/2 plywood
2 tray divisions	3	× 10 × 1/2 plywood

Rolling Tilt Bin

The bin in Fig. 4-11 is a container that tilts into its case, which is a table-high cabinet. When tilted out far enough to lift away, the bin can be rolled about on the floor. If you use it for trash in the kitchen or shop, you will be able to wheel it away instead of having to lift what might be a heavy load. Although shown as a self-contained unit, the same idea could be used under a kitchen counter.

The cabinet and bin front are 3/4-inch plywood. The other parts of the bin are 1/2-inch plywood. Strengthening pieces are 3/4-inch square strips. The wheels are 2-inch diameter, which you can buy or make yourself on a lathe. An alternative to wheels is to add a roller across. A 1/4-inch steel rod will serve as an axle.

The front of the bin hooks over a two-part crosspiece at the bottom of the cabinet (Fig. 4-12A). Each side is cut so that the bin bottom comes above the cabinet crosspiece, then slopes down to within 1/4 inch of the bottom of the cabinet (Fig. 4-12B). The wheels or roller then fit on their axle through the sides, so they clear the bottom of the bin (Fig. 4-12C).

At the top, the bin front closes against a similar two-part crosspiece (Fig. 4-12D). There is no need for a fastener as the bin will stay closed under its own weight.

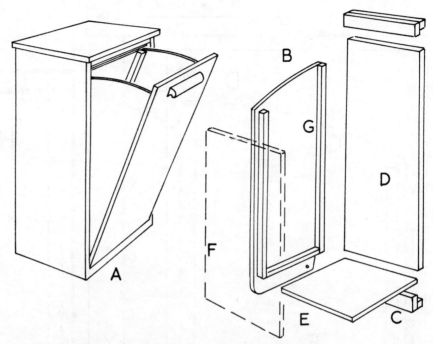

Fig. 4-11. This tilt bin rolls on wheels, and its bin can be lifted out.

1. The cabinet is boxlike, with an overhanging top and an open bottom. Make the two sides 14 inches × 29 1/4 inches. Make the back the same length and 11 1/2 inches wide. Make the front crosspieces from 1-inch and 2-inch pieces of 3/4-inch plywood 11 1/2 inches long. Cut the top 15 inches square.
2. Join all the cabinet parts with 5/16-inch dowels at about 4-inch centers.
3. Set out the bin sides (Figs. 4-11B and 4-12E). Sides come against the 3/4-inch front. At the back, sides should clear the inside of the cabinet by 1/4 inch. Cut the sides to hook over the crosspiece (Figs. 4-11C and 4-12B). Bevel slightly and round the wood so they clear the crosspiece when tilted. Round rear corners (Fig. 4-12F).
4. The top edge of the side has to clear the cabinet crosspiece as the bin is swung out. To draw this curve (Fig. 4-12G), use the bottom front corner as a center and draw a curve from this to clear the crosspiece. You could use an awl through a strip of wood and a pencil against the other end as a compass.
5. Make the bin front/door (Fig. 4-11D) to fit easily in the cabinet front opening.
6. The bin bottom (Fig. 4-11E) fits above the level of the front notches in the sides. The back (Fig. 4-11F) fits between the sides and above the bottom. Round its top edge.

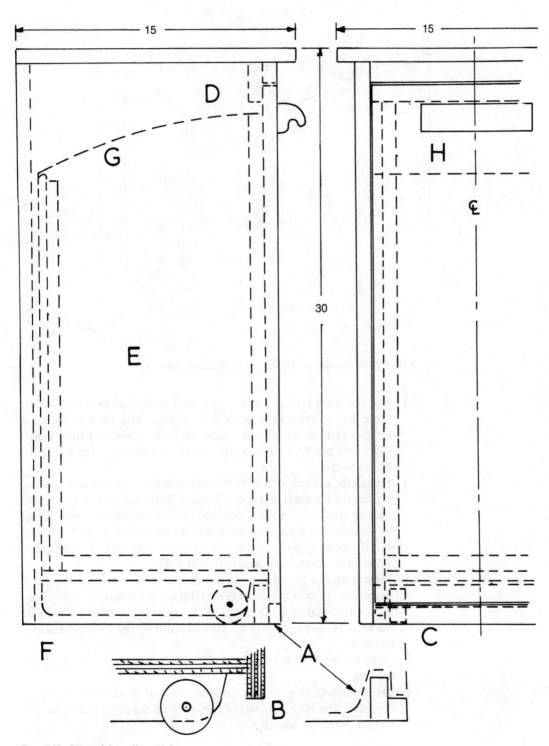

Fig. 4-12. Sizes of the rolling tilt bin.

7. Put 3/4-inch square strips on the sides to suit the positions of the other parts (Fig. 4-11G), using glue and pins.
8. Assemble the bin parts after checking their sizes in relation to the cabinet.
9. Add a handle near the top of the front. You could make or buy one. It might be a knob, but a strip handle is shown in Fig. 4-12H.
10. Fit the wheels or roller and the axle. Try the action, then remove sharp edges and corners before finishing with paint.

Materials List for Rolling Tilt Bin

2 cabinet sides	14	× 30 × 3/4 plywood
1 cabinet back	11 1/2	× 30 × 3/4 plywood
2 crosspieces	2	× 12 × 3/4 plywood
2 crosspieces	1	× 12 × 3/4 plywood
1 cabinet top	15	× 15 × 3/4 plywood
1 bin front	11 1/2	× 28 × 3/4 plywood
2 bin sides	12	× 28 × 1/2 plywood
1 bin bottom	12	× 12 × 1/2 plywood
1 bin back	11	× 23 × 1/2 plywood
4 strips	3/4	× 3/4 × 24
4 strips	3/4	× 3/4 × 12

Tabletop Containers

Boxes and trays can be used just about anywhere to hold desk supplies, cosmetics and toiletries, hobby equipment, or any type of small items. The containers in Fig. 4-13 are intended to be made from 1/4-inch plywood, preferably hardwood. They can be finished with paint or given a clear finish.

Sizes will depend on your needs, but suggested depths are given in Fig. 4-14A and B. Dimensions the other way could be anything from perhaps 3 inches square up to quite a large tray or box. If there are pens, tools, or other things of fixed size, make the containers to suit. For such things as paper clips, sizes can be almost anything.

You might want to make a set, so consider relative sizes and appearance. Two concepts are suggested. A tray (Fig. 4-13A) or box (Fig. 4-13B) might have straight edges and shaped ends. If you want more decoration, you can shape the edges (Fig. 4-13C and D). All joints are glued and secured with fine pins.

1. For the first tray in Fig. 4-13A, fit the bottom inside. Mark opposite sides together to get them the same (Fig. 4-14A). Notch halfway through and round the extending ends (Fig. 4-14C).
2. Join these parts, then make the bottom to fit tightly inside. Glue and pin it in.

96 *Containers*

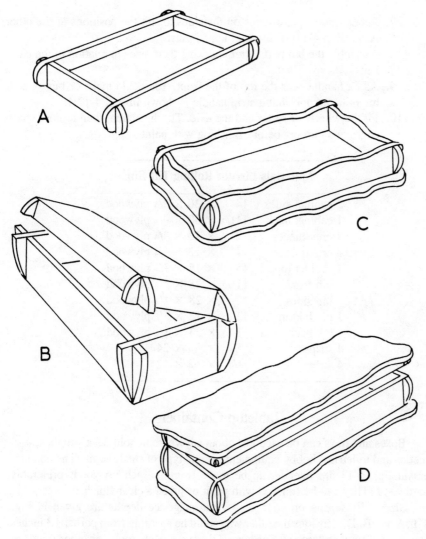

Fig. 4-13. These small containers are made with notched plywood sides.

3. The box in Fig. 4-13B is similar to the tray, with top and bottom fitted inside (Fig. 4-14B). Mark the lid and box parts together (Fig. 4-14D), so they will match. Shape the ends, then separate the lid from the box and cut the notches. Mark matching parts so they are not changed around or turned over during assembly.
4. Join the outside parts, then fit in the top and bottom with glue and pins. Add small decorative hinges on the rear surface if you like.
5. If you want to decorate with deckle edges as in Fig. 4-13C and D, fit the sides together in the same way, but fit the bottoms and tops on the outside and extend (Fig. 4-14E and F).

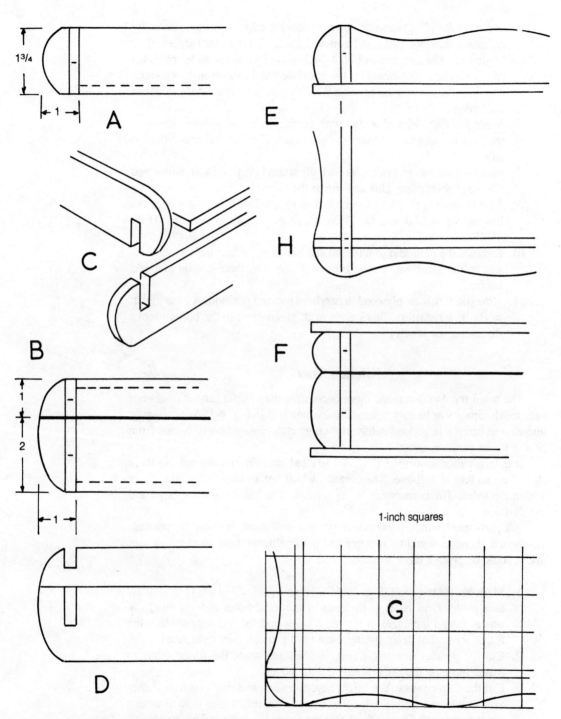

Fig. 4-14. Sizes and shapes of the tabletop containers.

98 Containers

6. For the sake of appearance, give the deckle edges uniform curves by making a template that can be moved along. It could be cardboard or hardboard. The template in Fig. 4-14G suits edges arranged in multiples of 2-inch steps. For example, the template would suit 6-inch-×-8-inch edges, but if you wanted to use a 9-inch edge, you would have to adjust the curves.
7. Make the four sides of a tray with curved ends and notched corners, then use the template to mark the top edges. Cut this and round the top edges.
8. Make the bottom to extend 3/4 inch all around (Fig. 4-14H). Mark and cut the shaping, then glue and pin to the sides.
9. Start by making the sides for the box in Fig. 4-13D the same way as the box in Fig. 4-14B and D. Different shaping is suggested here (Fig. 4-14F).
10. Separate the parts and join the lid and box sides. Make the top and bottom in the same way as the bottom of the tray, then join on and add hinges.
11. How you finish the plywood depends on use and position. A clear finish might be satisfactory. Black paint with green cloth in the bottom looks distinctive for jewelry.

Kitchen Tray

The usual tray is often made more decorative than useful, and it might not have much capacity or be easy to use. The kitchen tray in Fig. 4-15A is primarily intended to carry a large load safely and comfortably. Any beauty comes from fitness for its purpose.

It is larger than most trays (Fig. 4-15B) and soundly constructed, so there should be no fear of collapse. The bottom, which has to take the weight, is set within the sides. The corners are finger-jointed. The hand holes are large and comfortable.

All parts are 1/2-inch plywood. If you use softwood, it could be painted. Hardwood plywood would be stronger and better able to stand up to rough use, and it could be given a more attractive clear finish.

1. Mark out the two ends (Fig. 4-15C), using a grid of squares (Fig. 4-15D) as a guide. Drill holes at the ends of the hand holes and cut away the waste. Round the edges of the hand holes and the top edges. Mark the finger joints, but delay cutting them until you can match the sides.
2. Cut the parallel side strips (Fig. 4-15E) and mark the finger joints to match those on the ends. Round the top edges.
3. Cut the finger joints. You might have to match and mark corners if they are not interchangeable. Glue the corners. You might also wish to drive a pin through each finger. Check that the frame is square and without twist.
4. Fit 1/2-inch square strips all around inside the lower edges (Fig. 4-15F).

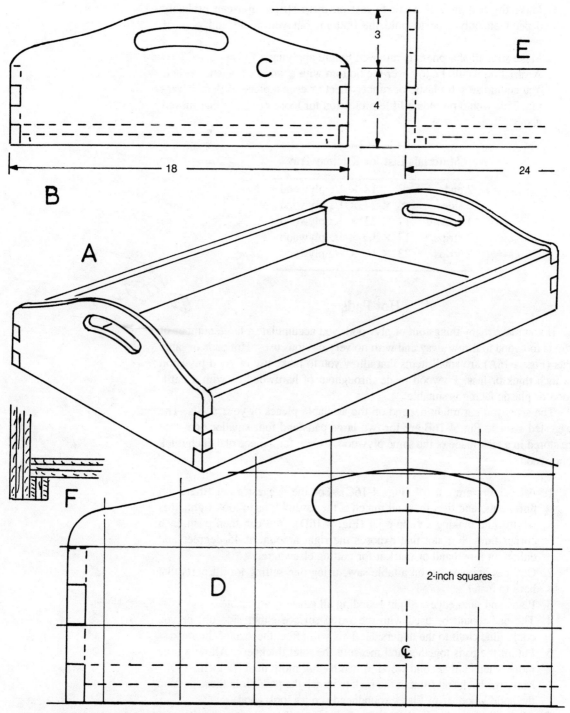

Fig. 4-15. This strong tray is intended for kitchen use.

Containers

5. Make the bottom a close fit to rest on these strips. You can probably depend on only glue to hold the bottom, but you could add pins, if needed.
6. Make sure all sharpness is removed before applying a finish.
7. A variation would be to cover the bottom with a wood or plastic veneer. You could use soft plastic or rubber sheet or even a piece of floor covering. This would provide a little protection for loose crockery that moved about.

Materials List for Kitchen Tray

2 ends	7 × 18 × 1/2 plywood
2 sides	4 × 24 × 1/2 plywood
1 bottom	17 × 23 × 1/2 plywood
2 strips	17 × 1/2 × 1/2 plywood
2 strips	23 × 1/2 × 1/2 plywood

Hot Pads

If you make many things out of plywood, you accumulate a large number of offcuts too good to throw away and with no very obvious uses. Hot pads or table mats (Fig. 4-16A) are small items that allow you to make use of good plywood 1/4 inch thick or less. Plywood made throughout of hardwood or with a hardwood or plastic facing is suitable.

The sizes you cut might depend on the available pieces or your needs. The suggested sizes in Fig. 4-16B are for two large pads and four smaller ones that are stored in a rack made of the same plywood, except for a piece of 1-inch thick solid wood.

1. All corners are cut off (Fig. 4-16C). For the suggested set, there are thirty cuts, and they must all match if the assembly is to look right. It is worthwhile making a simple jig (Fig. 4-16D). You can then push each corner through a slot that exposes the right amount at 45 degrees and either cut it by hand or mark it for cutting elsewhere.
2. Outlines are best cut on a table saw, using one setting for all parts that have to match.
3. Plane and sand edges. Slight rounding all edges is advisable.
4. The pads could be used with the wood surface either side up, or you could glue cloth to the underside. This will affect the size of the rack.
5. Put all the pads together and measure the total thickness. Allow a little more when cutting solid wood to width—not more than 1/8 inch.
6. Make the back and front of the rack to match the pads. Join these parts to the solid wood strip. Plated roundhead screws look good.
7. Mount these parts on a base extending 1 1/2 inches all around, with corners trimmed to match.

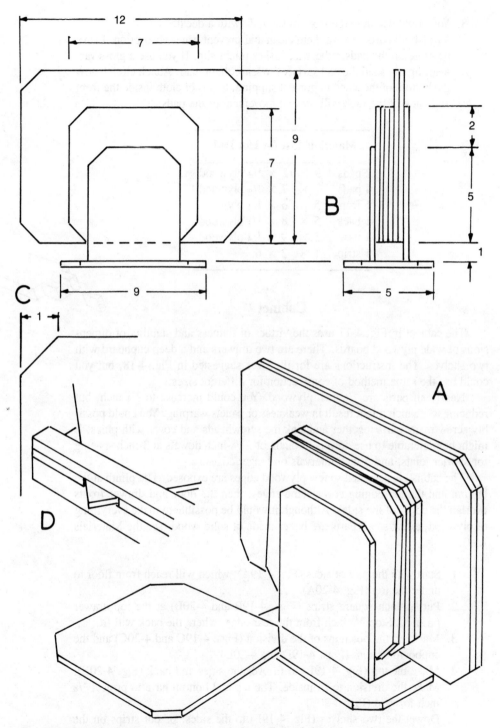

Fig. 4-16. Hot pads and their stand can be made from offcuts of plywood.

8. You could decorate the front of the rack with a decal.
9. Varnish all parts to keep them clean and prevent water absorption. Leave no gloss on the pads, otherwise dishes might slip. If you use a gloss varnish, lightly sand the pad surfaces to leave them matt. Attach cloth below the bottom of the stand to prevent slipping. Strips of cloth inside the front and back of the rack will prevent movement of the pads.

Materials List for Hot Pads

2 pads	$9 \times 12 \times 1/4$	plywood
4 pads	$7 \times 7 \times 1/4$	plywood
1 front	$5 \times 6 \times 1/4$	plywood
1 back	$5 \times 8 \times 1/4$	plywood
1 base	$5 \times 9 \times 1/4$	plywood
1 strip	$1 \times 2 \times 6$	

Cabinet

The cabinet in Fig. 4-17 uses the virtues of flatness and stability of dimensions of wide plywood boards. There are two drawers and a deep cupboard with two shelves. The instructions are for the sizes suggested in Fig. 4-18, but you could use the same method of construction for different sizes.

Nearly all parts are $5/8$-inch plywood. You could increase to $3/4$ inch, but reducing to $1/2$ inch might result in weakness or panels warping. You could possibly screw many parts together and sink the screwheads and cover with plugs. It might be preferable to use dowels throughout—$1/4$-inch dowels at 3-inch spacing for shorter joints, and 5-inch intervals on longer edges.

The cabinet is designed so few plywood edges are exposed. The plinth at the bottom and the top edging cover some edges, then the door and drawer fronts overlap the edges of the sides. Although it would be possible to make everything of plywood, a few small parts are better made of solid wood (see the Materials List).

1. Start with the pair of sides (Fig. 4-19A), which will reach from floor to under the top (Fig. 4-20A).
2. Put $5/8$-inch square strips (Figs. 4-19B and 4-20B) at the top drawer position. Stop $5/8$ inch from the rear edge, where the back will fit.
3. Mark out the positions of the division (Figs. 4-19C and 4-20C) and the cupboard bottom (Figs. 4-19D and 4-20D).
4. Make the top (Fig. 4-19E) to fit over the sides and back (Fig. 4-20E). Make the division to fit inside. The top and bottom have to project $5/8$ inch forward (Fig. 4-20F).
5. Dowel the two shelves (Fig. 4-19F) to the sides, or put strips on the sides for the shelves to rest on so that they can be lifted out.

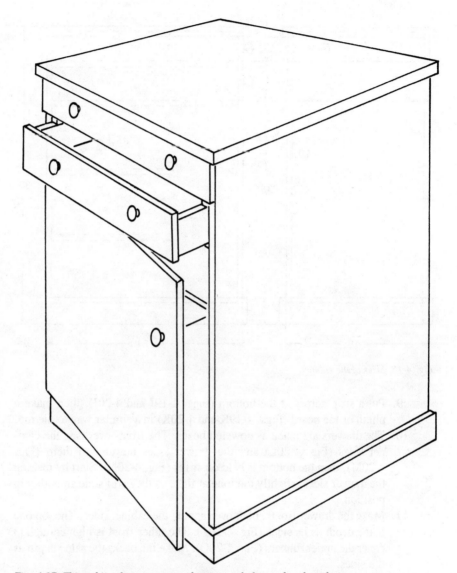

Fig. 4-17. This cabinet has two roomy drawers and plenty of cupboard space.

6. Make the back (Fig. 4-19G) to fit between the sides. It need not reach the floor but can stop below the cupboard bottom.
7. Prepare all these parts for dowel joints, then join them together. The back should keep the assembly square, but put the parts together on a flat surface and see that there is no twist.
8. Put a strip across at the top level with the side ones (Fig. 4-20G), then frame around the sides and front with strips (Figs. 4-19H and 4-20H) mitered at the corners. Round the outer corners of the strips. Glue these parts and hold them with pins set below the surface and covered with stopping.

104 *Containers*

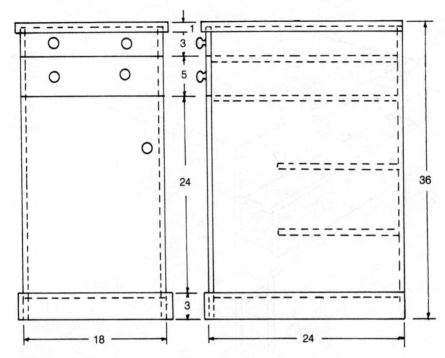

Fig. 4-18. Sizes of the cabinet.

9. Put a strip across at the bottom (Figs. 4-19J and 4-20J), then make a plinth or toe board (Figs. 4-19K and 4-20K) in a similar way to the top.
10. The drawers are made as doweled boxes. The fronts overhang the cabinet sides (Fig. 4-20L) and the drawer sides dowel into them (Fig. 4-20M), then the bottom is held on strips (Fig. 4-20N). Start by making the drawer sides, slightly too long at first, so they will slide smoothly in position.
11. Make the drawer fronts, both overlapping the cabinet sides. The top one is the depth as its sides (Fig. 4-20P). The other front is high enough to cover the upper runners (Fig. 4-20Q). Make the backs the same depth as the sides.
12. Join these drawer parts and check their action. Fit the inside strips and bottoms (Fig. 4-20N).
13. The door overlaps both sides and the division (Fig. 4-20R).
14. You could use ordinary hinges on the door, but the concealed throw-clear type are appropriate. Fit a catch and handle or knob, with matching handles on the drawers.
15. Finish with paint or varnish. A different color for knobs and top and bottom edges would look good. Even if ordinary plywood is used for other parts, a special veneer on the top or on drawer and door fronts would improve appearance.

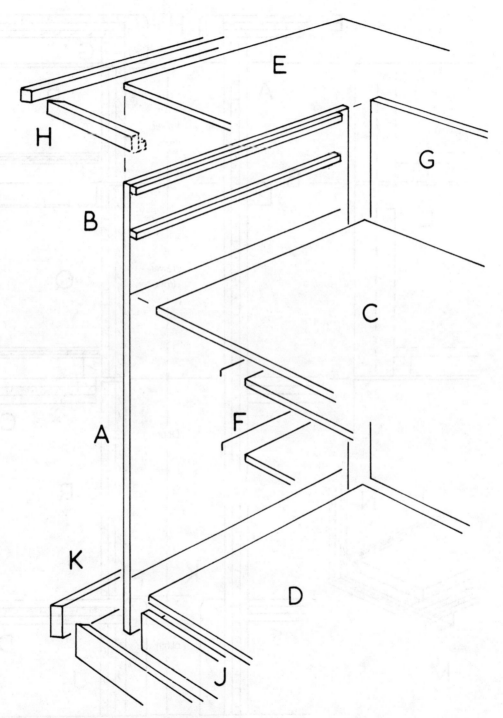

Fig. 4-19. How the cabinet parts fit together.

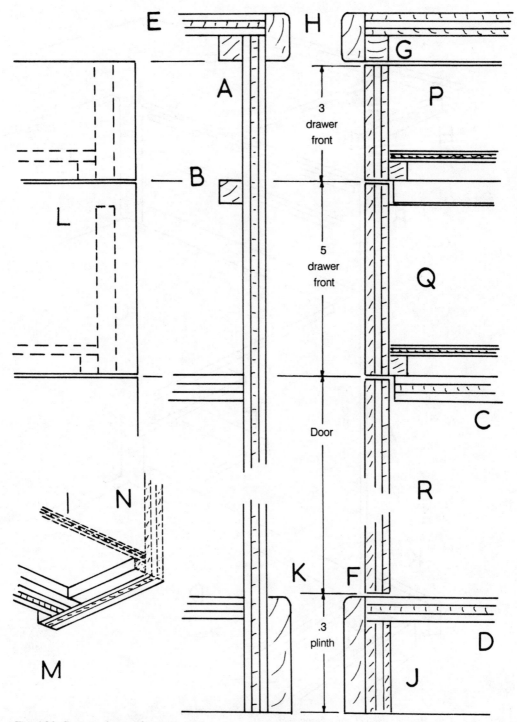

Fig. 4-20. Sections showing how parts of the cabinet are constructed.

Materials List for Cabinet

2 sides	24	× 36	× 5/8 plywood
1 top	19	× 25	× 5/8 plywood
1 division	18	× 24	× 5/8 plywood
1 bottom	19	× 24	× 5/8 plywood
1 back	24	× 36	× 5/8 plywood
1 door	24	× 24	× 5/8 plywood
2 shelves	16	× 18	× 5/8 plywood
1 drawer front	3	× 18	× 5/8 plywood
1 drawer back	3	× 17	× 5/8 plywood
2 drawer sides	3	× 24	× 5/8 plywood
1 drawer front	5	× 18	× 5/8 plywood
1 drawer back	4 1/2	× 18	× 5/8 plywood
2 drawer sides	4 1/2	× 24	× 5/8 plywood
6 strips	5/8 ×	5/8	× 24
2 strips	5/8 ×	1 1/4	× 26
1 strip	5/8 ×	1 1/4	× 19
2 strips	5/8 ×	3	× 19
2 strips	5/8 ×	3/4	× 26
2 drawer bottoms	18	× 24	× 1/4 plywood
8 drawer bottom supports	1/2 ×	1/2	× 24

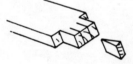

Box/Roll Holder

This is a wall-mounting storage box with a rail below for paper towels that you can use in the kitchen, although it could find a place in a garage, your garden store, or anywhere else you want to store small items or use a paper towel (Fig. 4-21). As drawn in Fig. 4-22A, the project will hold any paper roll up to 11 inches long and 6 inches in diameter on a 3/4-inch rod. You can easily modify sizes if you want to accommodate a different roll.

All parts are 1/2-inch plywood. You could glue and nail or screw them, although dowels no thicker than 3/16 inch might be used if you want to avoid heads showing outside. All crosswise parts come between the sides, which should be made first.

1. Set out the two sides (Figs. 4-22A and 4-23A) with the positions of the other parts marked on them. Drill to suit a 3/4-inch dowel rod. Cut and round the shaped edges.
2. Make the back the same height as the sides and to fit between them. Shape the top (Fig. 4-23B). You could shape the bottom similarly, but it will be hidden by the paper towels, so you might decide to leave it straight. Use the width of the back as a guide when you cut the other parts, so their lengths match.

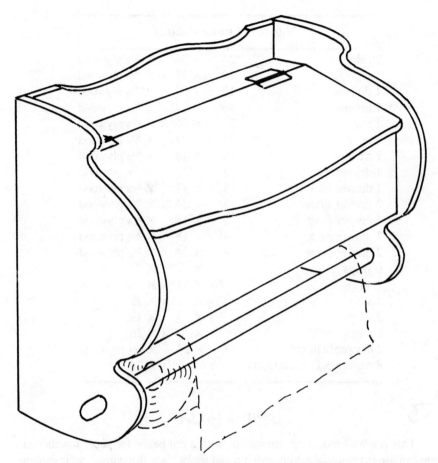

Fig. 4-21. This wall-hanging box has a rod for a paper roll underneath.

3. Cut the front (Fig. 4-22B) and the bottom (Fig. 4-22C) to fit behind it. Use the grid drawing as a guide to the angle of the top edge of the front (Fig. 4-23C).
4. Make the top (Fig. 4-22D) and the lid (Fig. 4-22E). Bevel them to match where they meet (Fig. 4-23D) and shape the front of the lid (Fig. 4-23E). Round that edge.
5. Join these parts between the sides and hinge the lid along its rear edge. If there is any risk of dampness, use brass hinges.
6. The rod (Fig. 4-22F) should project at the ends, for ease in pushing it out. Doming the ends in a lathe will improve appearance.
7. The best finish will probably be paint, possibly with a lighter color inside than out.

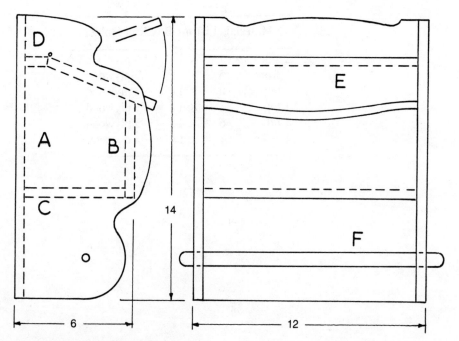

Fig. 4-22. Sizes of the box/roll holder.

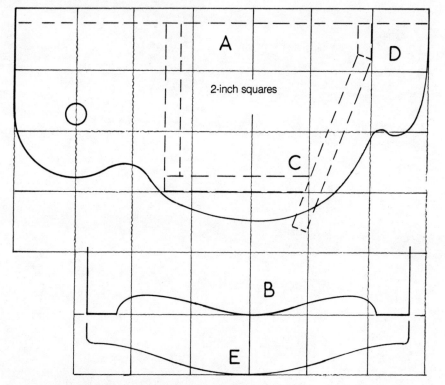

Fig. 4-23. The shaped parts of the box/roll holder.

Materials List for Box/Roll Holder

2 sides	7	× 14 ×	1/2 plywood
1 front	5	× 11 ×	1/2 plywood
1 bottom	5	× 11 ×	1/2 plywood
1 top	1 1/4	× 11 ×	1/2 plywood
1 lid	6	× 11 ×	1/2 plywood
1 rod		14 ×	3/4 diameter

5

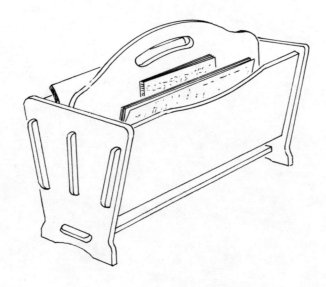

Racks and Shelves

Souvenir Rack

The souvenirs and mementos you accumulate from trips and vacations are often small, and they ought to be displayed. The rack in Fig. 5-1 is intended to hang on a wall and provide space for many small ceramic, metal, or wood items,

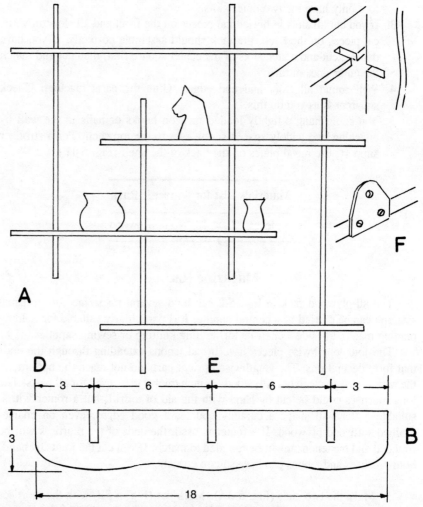

Fig. 5-1. This hanging souvenir rack is made from slotted strips.

114   Racks and Shelves

while also being decorative in itself. It has slotted joints, which you will probably glue, but you could leave the joints dry to reduce the rack to eight flat pieces if you want to store it.

The design shown in Fig. 5-1A is based on 6-inch squares with shelves 3 inches wide, but you could use other sizes to suit your needs. Front edges could be straight, but light scalloping and rounding improves appearance (Fig. 5-1B). All the parts fit into each other with slots (Fig. 5-1C).

1. Make eight strips to the outline shown (Fig. 5-1D). Lightly mark the positions of the outer slots on the strips, drawing the lines right across. On four, only mark the positions of the middle slots (Fig. 5-1E) in the same way.
2. The slots have to be cut so they alternate back and front. Carefully mark and cut the slots to keep the edges level. Four pieces have three slots and four only have the two outer slots.
3. If you cut all slots in horizontal pieces on the front and all slots on vertical pieces on the back, the rack should assemble correctly. If you have doubts, cut and match slots of the center square first, then cut and match the outer ones in turn.
4. Well round all front and end edges. Glue the parts together. Check squareness as you do this.
5. You could hang a lightly loaded rack on hooks or nails in the wall by attaching two widely spaced screw eyes to the top shelf. For a stronger support, use metal plates on the back of the shelf (Fig. 5-1F).

Materials List for Souvenir Rack

---
8 pieces   3 × 18 × 3/8 plywood

Magazine Rack

The all-plywood rack in Fig. 5-2 can hold several magazines of the usual size and can be carried by a central handle. It is particularly suitable for holding reading material beside a chair or for holding knitting or sewing supplies.

The four lengthwise pieces have broad tenons extending through the ends that form decorations. The lengthwise upright parts do not reach the bottom, so the rack is easy to clean out. Plywood 1/2 inch thick can be used for all parts. The long mortises could be cut by hand with the aid of a drill, but a router with a suitable cutter will shape accurately and leave good edges, even on coarse-grained softwood plywood. If a router is used, the ends of the mortises will be rounded and the tenons might be rounded to match. If you cut the joints by hand, both mortises and tenons can have square corners.

1. Make the two ends (Fig. 5-3A). Mark out (Fig. 5-3B) with the mortises extended to show the width of the joining parts. Cut the mortises before

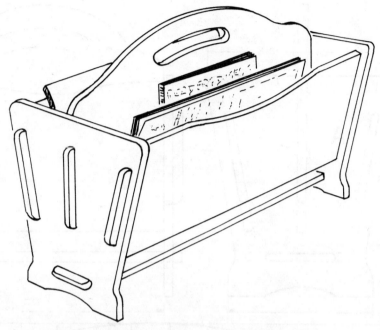

Fig. 5-2. The lengthwise parts of this magazine rack tenon into the ends.

trimming the outside to size; this reduces the risk of veneer edges breaking out.

2. Mark out the central handled part (Fig. 5-3C). Shape the top and hand hole (Fig. 5-3D). Cut the tenons to go through the ends and extend 1/2 inch with rounded corners (Fig. 5-3E).
3. The two sides (Fig. 5-3F) must be the same length between shoulders as the central part and their tenons extend the same. Check their depth as marked on the end pieces. Hollow out the top edges, if you wish.
4. Make the bottom the same way as the sides. Round the long edges of all lengthwise parts.
5. After a partial trial assembly, sand all parts. A tight trial assembly is usually inadvisable, as this might result in the final joints being loose.
6. Glue the joints and draw them tight. Check that there is no twist in the length.
7. Finish with paint or varnish.

Materials List for Magazine Rack	
2 ends	11 × 12 × 1/2 plywood
1 center	12 × 16 × 1/2 plywood
2 sides	9 × 16 × 1/2 plywood
1 bottom	6 × 16 × 1/2 plywood

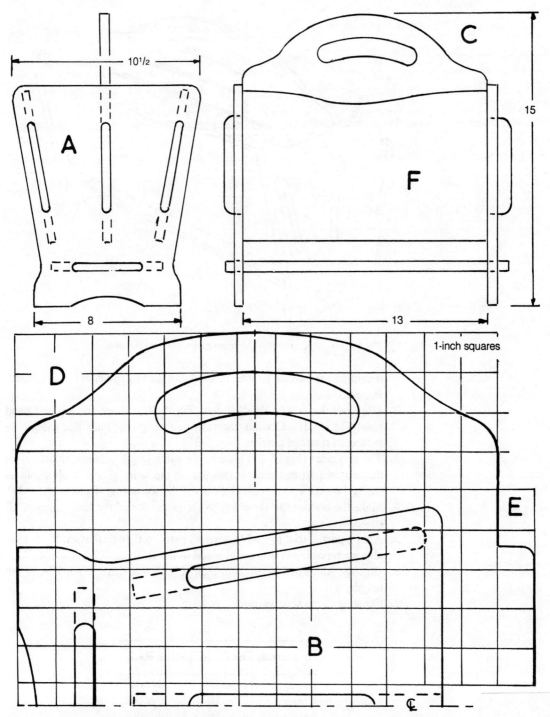

Fig. 5-3. Sizes and shapes of the magazine rack.

Take-down Bookshelves

The hanging block of shelves in Fig. 5-4 can be taken apart and stored or transported flat. The shelves have tenons through the ends that are secured by wedges made from dowel rods in a similar way to the tusk tenons of Tudor furniture. Various sizes are possible, but the example will take two rows of books up to the size of this book you are reading, and there is space on top for other items.

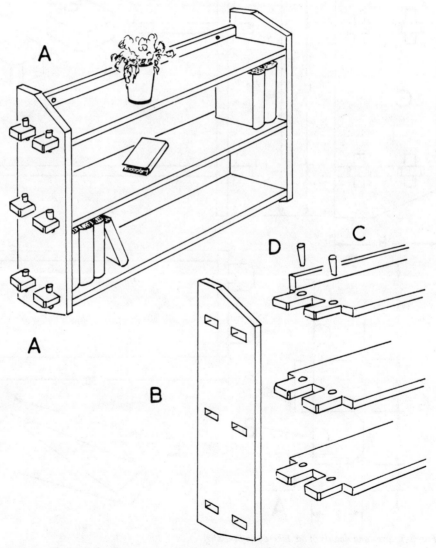

Fig. 5-4. This hanging bookcase can be taken apart and packed flat.

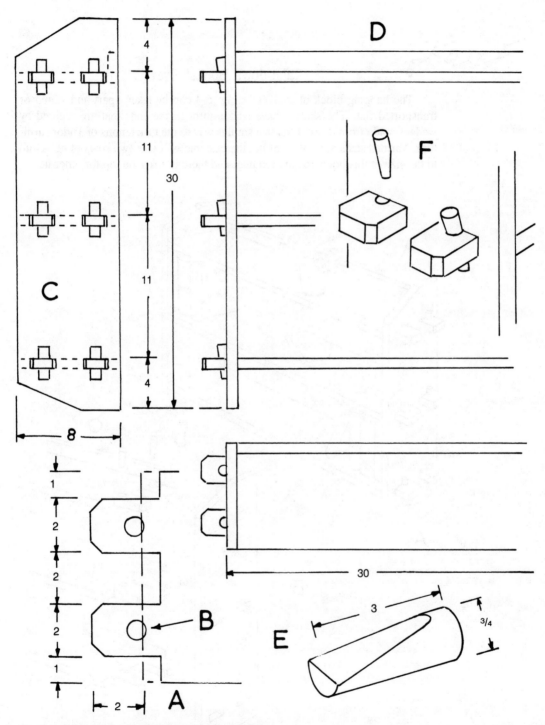

Fig. 5-5. Sizes and details of the take-down bookshelves.

The plywood must be stiff enough to resist sagging under a load of books. Some 1/2-inch plywood might be satisfactory, but 3/4-inch plywood is suggested. The wedges are made from 3/4-inch diameter hardwood dowel rod.

1. Make the three shelves exactly the same length between shoulders, then allow for the thickness of the plywood and a further 2 inches at each end (Fig. 5-5A).
2. Mark the tenons and the positions of the 3/4-inch holes, which should overlap the thickness of the ends by 1/8 inch (Fig. 5-5B).
3. Mark out the two ends (Fig. 5-5C) with mortises to match the tenons on the shelves (Fig. 5-4B).
4. Cut the mortises and make the tenons to match them. Ideally, the shelves will be interchangeable, but there can be slight differences. You will have to mark the positions for assembly. Bevel the outer corners of the tenons and drill the holes. Shape the top and bottom of each end.
5. Fit the hanging strip to the top shelf (Figs. 5-4C and 5-5D). This could be plywood, but solid wood might be better. Glue and screw the strip to the shelf, not to the ends.
6. Make the twelve wedges (Figs. 5-4D and 5-5E) the same. If you have a lathe, you could improve appearance by rounding the ends. Bevel from near one end to half the thickness at the other end, though you might wish to experiment with the amount of bevel, so when a wedge is driven, it goes halfway through (Fig. 5-5F).
7. Take sharpness off all exposed edges. Make sure there is enough clearance for the tenons to enter the mortises after being painted or varnished.

Materials List for Take-down Bookshelves		
3 shelves	8 × 36	× 3/4 plywood
2 ends	8 × 32	× 3/4 plywood
1 hanging strip	3/4 × 1 1/2 × 30	
12 wedges	40 × 3/4 diameter	

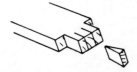

Vegetable Rack

The rack in Fig. 5-6 is designed for vegetables, but you might think of other things it could contain. It will serve as a stand in the kitchen, but the two hand holes allow you to carry it easily from an outside storeroom or to bring in your cooking needs from a bulk supply room. There is a large compartment at the bottom for such things as potatoes, plus two V-shaped troughs that can hold a variety of vegetables. The shallower tray at the top can hold small items, packages, and cans. The rack is stable, although it should be light enough to carry with a moderate load of vegetables.

120 Racks and Shelves

Fig. 5-6. You can carry this floor-standing vegetable rack.

All of the parts are ³/₄-inch plywood. Some, or all, parts are nailed or screwed, but for joining the horizontal parts to the uprights, ⁵/₁₆-inch dowels are suggested.

1. Start by marking out a side (Fig. 5-7A) with the positions of other parts shown. The bottom compartment is ¹/₂ inch in from the edges. The two V-shaped troughs are at a 45-degree angle to the centerline. The hand hole is 1 inch × 4 inches. Allow for a hollow on the bottom edge, so the rack will stand on its corners.
2. Mark the second side from the first and see that they match. Make the hand holes by drilling the ends and removing the waste between. Well round the hole edges and all of the outside of each top.
3. It is important that all of the crosswise pieces be the same length. You might be able to cut all at one setting of a table saw. Otherwise, you should make one piece and use it as a template for marking the others.

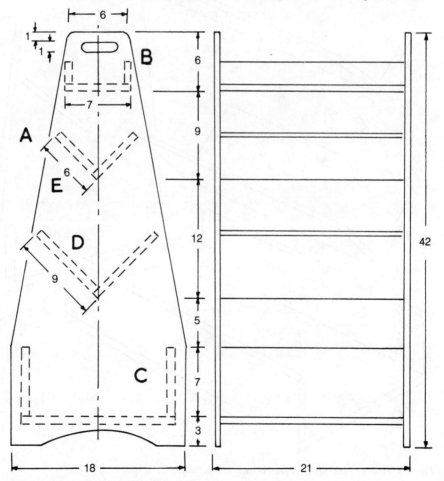

Fig. 5-7. Sizes of the vegetable rack.

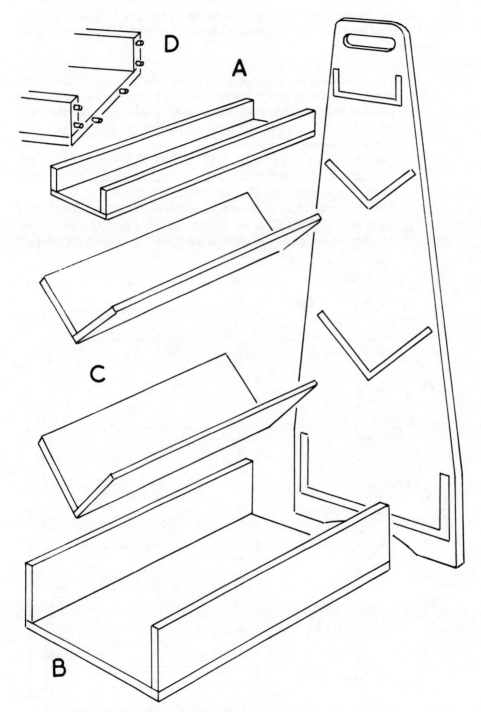

Fig. 5-8. How the parts of the vegetable rack fit together.

4. Make the top shallow tray (Figs. 5-7B and 5-8A) and glue and nail or screw the bottom to the sides. Round the top edges.
5. Make the bottom compartment (Figs. 5-7C and 5-8B) in the same way.
6. The two troughs are different widths (Fig. 5-7D and E) but are made with one piece overlapping the other, so allow for these laps when cutting the pieces to width (Fig. 5-8C). Round the outer edges and glue and nail or screw the joints.
7. Check the parts against their marked positions on the sides and mark the centers for dowels. Have at least two dowels in each crosswise end (Fig. 5-8D) and at no more than 3-inch intervals in wider pieces. Drill into the sides $1/2$ inch deep; the dowels can go $3/4$ inch into the crosswise ends. Drill all the parts and cut sufficient dowels so that you can complete the assembly at one time. It is possible to make all the joints first at one side only, then wait to join the other side, but it is easier to get the assembly square and tightly clamped if you make all the joints at once.
8. An alternative to dowels is to drive screws from outside. If you do not want the screwheads to show, counterbore the holes for plugs over the screwheads.
9. A painted finish will probably be best, but use a paint that is unaffected by moisture and is safe with food.

Materials List for Vegetable Rack

2 sides	18	× 42 × $3/4$	plywood
1 tray bottom	7	× 21 × $3/4$	plywood
2 tray sides	2	× 21 × $3/4$	plywood
1 tray bottom	17	× 21 × $3/4$	plywood
2 tray sides	7	× 21 × $3/4$	plywood
1 trough side	9	× 21 × $3/4$	plywood
1 trough side	$8 1/4$	× 21 × $3/4$	plywood
1 trough side	7	× 21 × $3/4$	plywood
1 trough side	$6 1/4$	× 21 × $3/4$	plywood

Hat and Coat Rack

If you have a powered scroll saw, you will want to use it for this project. The hat and coat rack in Fig. 5-9A has plenty of shaped edges only a scroll saw can cut easily and neatly. The rack is intended to mount on the wall and has three coat hooks and a shelf for hats and gloves.

The rack is intended to be made of $1/2$-inch plywood and would look best in hardwood. The rear corners are shown cut square, but you could extend the back if you want more shaping (Fig. 5-9B). The shelf is tenoned into the ends, but the rear joints are glued and screwed—thin screws, such as #4g × $1 1/4$ inches long should be satisfactory in the edge of most plywood.

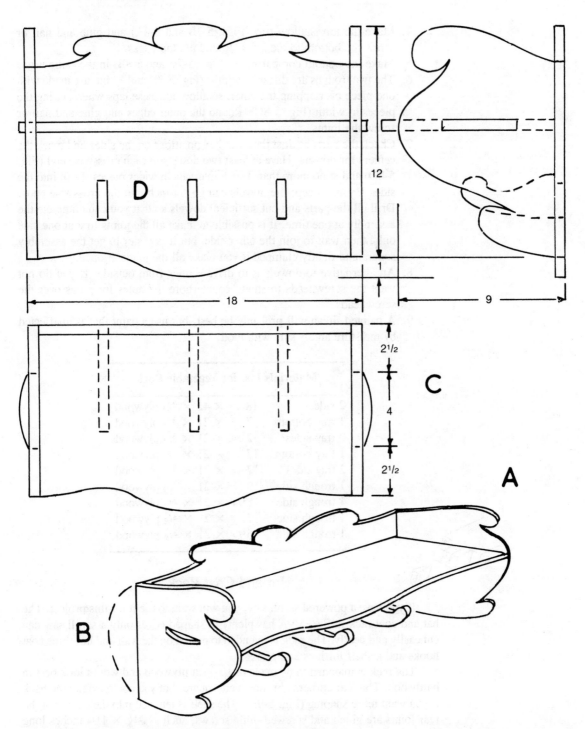

Fig. 5-9. This hanging hat and coat rack will appeal to the scroll saw user.

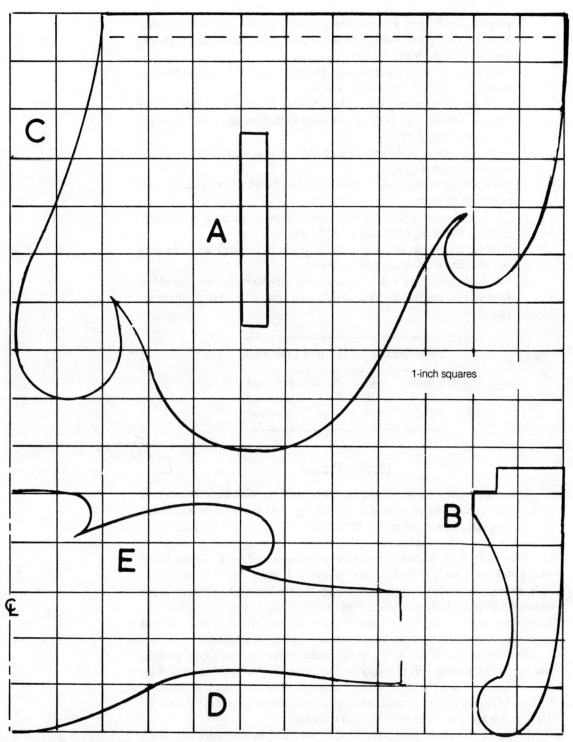

Fig. 5-10. The shaped parts of the hat and coat rack.

1. Prepare the plywood with the straight edges marked square and to size. Mark the shelf position on the back and ends. Mark where the coat hooks will come on the back.
2. Cut the tenons with curved ends on the shelf (Fig. 5-9C) and the mortises to match them in the ends (Fig. 5-10A).
3. Mark out the three coat hooks (Fig. 5-10B). Mark and cut mortises for them in the back (Fig. 5-9D). Cut the hooks to shape and round the edges of the projecting parts.
4. Mark the outlines of the ends (Fig. 5-10C). Cut to shape and remove any roughness from the edges.
5. Mark and cut the front edge of the shelf (Fig. 5-10D). Check that its ends match the rack ends.
6. The bottom edge of the back is the same shape as the front of the shelf. Also cut its top edge to shape (Fig. 5-10E).
7. Glue the coat hooks in place. Join the tenons to the mortises in the ends and glue and screw these parts to the back.
8. Two holes for screws widely spaced in the back above the shelf should be sufficient for hanging the rack. Apply your chosen finish to complete the rack.

Materials List for Hat and Coat Rack

1 back	13 × 19 × 1/2 plywood
1 shelf	10 × 20 × 1/2 plywood
2 ends	9 × 13 × 1/2 plywood

Picture Frame

This is a project for a scroll saw user, who will also find a use for a router with a cutter for beveling an edge. As drawn in Fig. 5-11A, the frame is intended for a photograph, picture, or mirror 10 inches × 12 inches, but the pattern can be used as a guide for other sizes.

You could use a softwood plywood and give it a painted finish, but the frame would look good made from hardwood plywood. If made with the same wood right through, with thin veneers to make up the thickness, the exposed edges trimmed with sharp tools and then sanded can be regarded as decorative features. More open wood for the inner veneers might tear out, but even then, painted edges around a choice hardwood front veneer would be distinctive.

You will have to decide on the space needed in the rabbets. Glass, picture, and hardboard backing will probably be contained in a 3/8-inch thickness (Fig. 5-11B). Brads or pins can angle through the edge of the hardboard. A mirror is better held in with a narrow strip at each edge, with card to protect the glass (Fig. 5-11C). For this you might need 1/2-inch thickness.

If you use 1/4-inch or thinner plywood for the front, its inner edge can be left

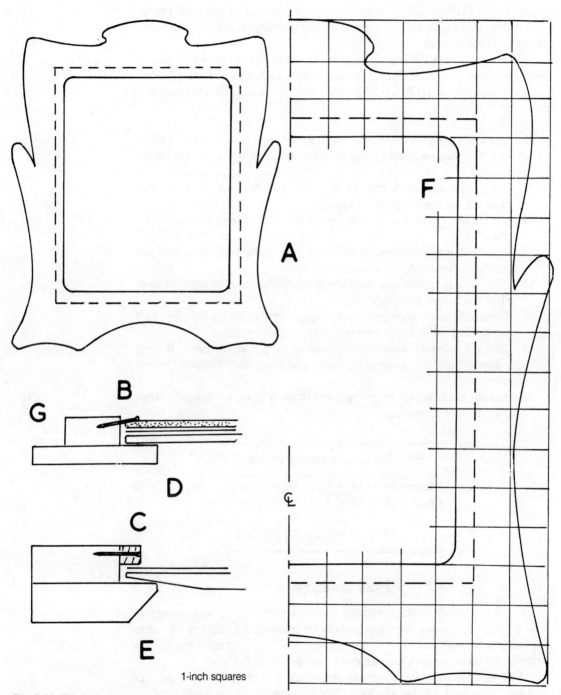

Fig. 5-11. *This frame for a mirror or picture can hang or stand.*

square (Fig. 5-11D) or lightly rounded. Thicker plywood on the front can be beveled (Fig. 5-11E). Corners of the opening can be rounded (Fig. 5-11F), so the router can follow around.

You can make both thicknesses of plywood the same outline, which gives a fairly strong appearance, or if you want the frame to look right, the rear plywood can be cut back (Fig. 5-11G). Its outside can then be parallel to the opening and 1 inch wide all around.

1. Take a piece of plywood large enough for the front and mark the opening on it. Cut this, preferably using the outer straight edges as a guide for the router.
2. In a similar way, mark and cut the opening in the back piece. On that piece cut the inner corners square.
3. Use the grid drawing as a guide for marking the outline on the front part (Fig. 5-11F).
4. If the back is to be trimmed to the same shape as the front, glue it on and cut the outline through the whole thickness.
5. If the back is to have a reduced straight outline, glue it on after the front piece has been cut to shape.
6. The frame can stand upright if you hinge a strut behind the top edge, or it can be hung from an eye or metal fitting screwed in place.
7. Sand and finish the wood before mounting the picture or mirror. If using a mirror, paint the inside of the rabbet black to reduce unwanted reflections.
8. You can keep dust out by gluing paper strips or using self-adhesive strips over the rabbet openings.

Materials List for Picture Frame
1 front   14 × 18 × 1/4 or 1/2 plywood
1 back   14 × 14 × 3/8 or 1/2 plywood
or
12 × 14 × 3/8 or 1/2 plywood

Floor Bookcase

If you have many books, a hanging bookcase might not be large enough to hold them all. In any case, books in quantity are heavy, and it might be unwise to depend on a few wall fastenings to take the weight. If there are many books, it is better to put them in a rack that stands on the floor (Fig. 5-12A).

A bookcase needs to be functional, but it could also be a presentable piece of furniture. This bookcase has spaces at three different depths for books, and the top shelf can take ornaments or more books. You might wish to measure your books and modify sizes to suit, if necessary. There are moderate decorative curves at the top and the fronts of the shelves.

Floor Bookcase 129

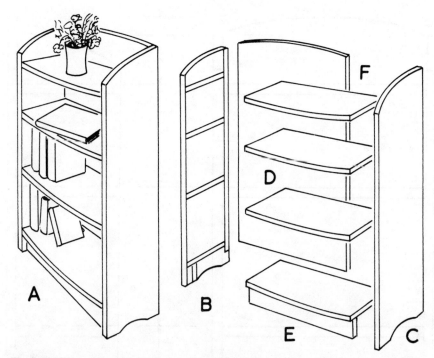

Fig. 5-12. You can store books of many sizes in this bookcase.

All parts can be ³/₄-inch plywood. The joints are made with ⁵/₁₆-inch dowels. The back could be thinner plywood, but all parts will cut from little more than half a standard sheet of ³/₄-inch plywood. The front and top edges could be covered with iron-on matching veneer, which will conform to the moderate curves without difficulty.

1. Mark out the sides (Figs. 5-12B and 5-13A) with the positions of other parts. The back need only reach to just below the bottom shelf. The toe board is set back 1 inch. Cut a curve to make feet from just behind the toe board (Fig. 5-12C).
2. Make the four shelves the same (Figs. 5-12D and 5-13B). Spring a 1-inch curve at the front. Fit the toe board below the bottom shelf (Fig. 5-12E).
3. The back (Fig. 5-12F) is the same width as the lengths of the shelves, and its top curve matches the fronts of the shelves.
4. Iron on self-adhesive veneer on all front and top edges.
5. Prepare the parts for dowels at about 3-inch intervals in the shelf ends (Fig. 5-13C). Between the back and the sides they could be 6 inches apart. You could use screws between the back and the shelves. Let the dowels go at least ¹/₂ inch into the sides.
6. Make and clamp all joints at the same time, if possible, but you could

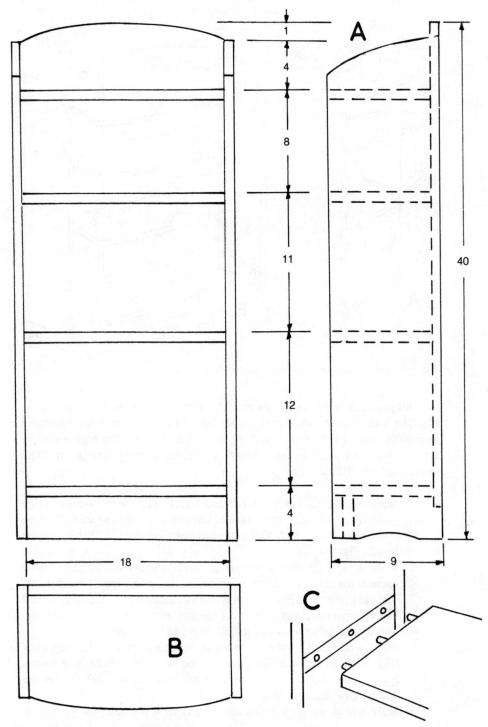

Fig. 5-13. Sizes of the floor bookcase.

wait to join the second side if it is more convenient to complete the other joints first.
7. If you have used a hardwood plywood, finish with varnish or polish. It might be better to paint softwood plywood.

Materials List for Floor Bookcase

2 sides	9	× 40 × 3/4 plywood
4 shelves	9	× 18 × 3/4 plywood
1 toe rail	3 1/4	× 18 × 3/4 plywood
1 back	16 1/2	× 40 × 3/4 plywood

Drafting or Macramé Desk

A board that can be mounted on an ordinary table will serve as a drafting desk when you want to lay out your own designs. Anyone interested in macramé or other craft, where the working surface needs to be supported conveniently, can use a similar board. This desk is arranged at about 30 degrees, but it can be folded flat (Fig. 5-14A). The size suggested is 24 inches × 36 inches, which will cope with many drawings, but for macramé alone, it need not be as big.

There are folding supports underneath (Fig. 5-14B). Below the front edge is an extending strip with two stops (Fig. 5-14C) to fit against the straight or curved edge of a table. For extra rigidity, you can use C clamps on the ends.

There is a ledge on the board (Fig. 5-14D), which will serve as a stop for drawing instruments or a place for clipping cords when doing macramé. For drawing, you will probably want a T square. For macramé, use a piece of softboard or insulation board for pinning cords.

Use hardwood plywood if possible—1/2 inch will probably be stiff enough. If you use softwood plywood it will need to be thicker for stiffness, and you must expect to do some thorough sanding to obtain a good working surface. The narrow parts could be plywood or solid wood strips.

1. Make the board first (Fig. 5-15A). Check that the chosen piece of plywood is flat. Plane and sand the edges straight and square. If you will do much drawing, it is worthwhile putting a solid wood lip on the left-hand edge so the stock of a T square will slide easily.
2. Put 1/2-inch-×-1-inch strips 3 inches in from the ends (Figs. 5-14E and 5-15B).
3. Cut the two supports (Fig. 5-15C) about 30 degrees, but do not attach them yet.
4. Make the front strip (Fig. 5-15D). Bevel it to the same angle as the supports and let it extend 3 inches each side of the board (Fig. 5-14F).
5. Put stop blocks under the strip (Fig. 5-15E).

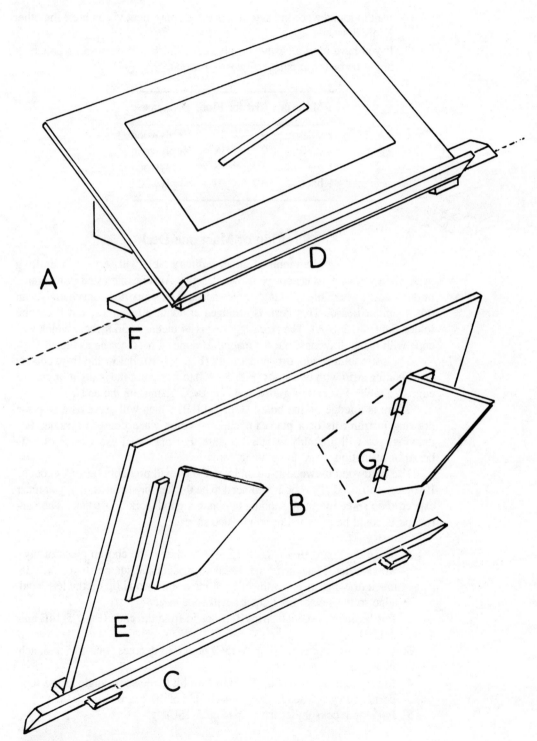

Fig. 5-14. A desk for drafting or macramé can fit over the edge of a table.

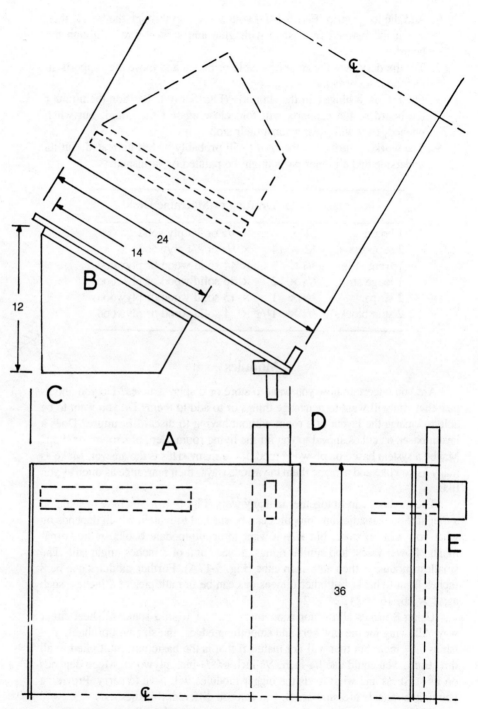

Fig. 5-15. Details and sizes of the drafting desk.

6. Add the ledge strip (Fig. 5-14D) with screws up through the board, then attach the beveled front strip with glue and screws down through the board.
7. Try the desk over the edge of a table or bench and move the supports so they hold it correctly.
8. Put 1 1/2-inch hinges on the supports (Fig. 5-14G). Position them under the board so the supports will fold close against the board, but when opened, they will press against their stop strips.
9. The working surface of the board will probably be left untreated, but its underside and all other parts might be painted or varnished.

Materials List for Drafting or Macramé Desk

1 board	24 × 36	× 1/2 or 3/4 plywood
2 supports	12 × 14	× 1/2 or 3/4 plywood
1 front strip	3/4 × 3	× 44 solid wood or plywood
1 ledge strip	3/4 × 2	× 36 solid wood or plywood
2 strips	1/2 × 1	× 15 solid wood or plywood
2 stop blocks	3/4 × 1 1/2	× 4 solid wood or plywood

Modules

Are you uncertain how you want to store or display articles? Do you anticipate that you will want to rearrange things or to add to them? Do you want to be able to change the layout of a room without having to discard furniture? Does a layout based on cubes appeal to you for the living room, den, playroom, or shop? Maybe a system based on plywood modules of many sizes is the answer. Make as many as you like and arrange them in various ways, then rearrange as often as you like.

The modules can fit together in many ways (Fig. 5-16). You have to settle on a unit size so that adjoining modules can be stacked to match. Much depends on what you wish to store, but if you want to accommodate books, radio, ornaments, flower vases, and similar things, a basic unit of 8 inches might suit. The smallest module is then an 8-inch cube (Fig. 5-17A). Further modules can be 8 inches front to back, but other dimensions can be in multiples of 8 inches, such as: 8 × 16, 16 × 24, 24 × 24.

Using 8 inches allows for economical cutting from a standard sheet either way. Allowing for the saw kerf and smoothing edges, the size will probably finish as 7 3/4 inch, but that will not matter if that is the basic unit multiplied in all directions. You could use 1/2-inch, 5/8-inch or 3/4-inch plywood. Much depends on its stiffness and what loads the bigger modules will have to carry. Providing you match outside measurements, you can use different thicknesses.

Although you could make modules one at a time, it will be more efficient and accurate to make your first set as a series. You can ensure uniformity by

Fig. 5-16. Modules can be assembled and rearranged to form many types of furniture.

using one setting of a table saw, for instance. Joints can be marked and cut together. The following instructions are for one module, but multiply these to suit your intended output.

1. Cut a set of four sides squarely.
2. Mark the finger joints, which could be at 1-inch spacing. Allow a little extra for trimming joint ends level after gluing (Fig. 5-17B).
3. Join the corners with glue. Clamp tightly and squarely. You could drive a pin into each leaf of the joint for extra strength.
4. Make a back (Fig. 5-17C) to fit inside. Fix it with glue and pins. For a simpler back, you could nail on hardboard. For some modules you might not need a back, such as against a wall or when you want access from both sides as in a room divider.
5. Level the joint ends and lightly round all edges, then finish with paint or varnish.

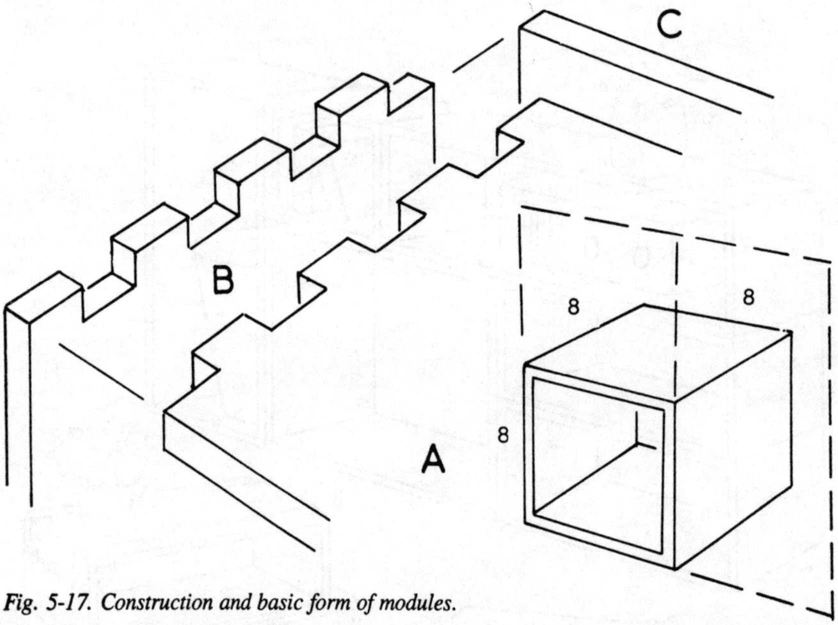

Fig. 5-17. Construction and basic form of modules.

6. You could prepare a few plain plywood boards, using the same 8-inch unit. One might stand on two modules as a table. You could include a board in a stack as a shelf over a space or to support several narrower modules above an opening.

6

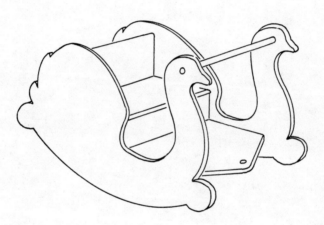

Toys

Doll's Crib

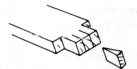

A rocking crib will appeal to any young girl. The one in Fig. 6-1 will provide space for bedding and a small doll, but the sizes can be altered if the favorite doll needs more space. For the sizes shown (Fig. 6-1A), most of the parts can be 1/4-inch plywood, with 1/2-inch square strips forming joints.

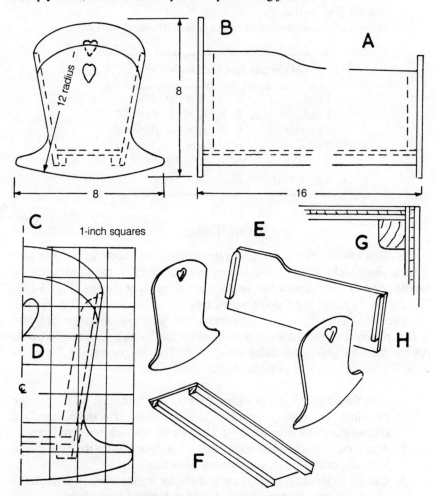

Fig. 6-1. A doll's crib follows traditional lines and is built of framed plywood.

1. Set out the high end (Fig. 6-1B and C). The rocker curve is a 12-inch radius. Mark the positions of the sides and bottom, which the ends will overlap. Mark out the low end from this. Its top is 1 inch lower, but it has a matching curve. A heart cutout is shown for the low end (Fig. 6-1D). You can make a similar one 1 inch higher on the other end.
2. The markings on the crib ends will show you the widths at the ends of the sides. Step down 1 inch with a curve about 5 inches from the high end (Fig. 6-1E). The sides overlap the bottom and its joint pieces. Round the top edges.
3. Make the bottom the same length as the sides. Get its width and angles from your setting out of the high end. Bevel joint pieces to match. Glue and pin them on (Fig. 6-1F).
4. Prepare side joint pieces (Fig. 6-1G) with rounded inner corners.
5. Glue and pin the bottom between the ends, then add the sides with the corner joint pieces, which should touch the bottom and have their tops rounded (Fig. 6-1H).
6. Remove any sharp angles or edges before finishing with paint.

Materials List for Doll's Crib

1 end	8 × 9 × 1/4 plywood
1 end	8 × 8 × 1/4 plywood
1 bottom	4 × 16 × 1/4 plywood
2 sides	6 × 16 × 1/4 plywood
2 joint strips	16 × 1/2 × 1/2
4 joint strips	6 × 1/2 × 1/2

Swan Rocker

A toddler likes a rocking seat, and the junior swan rocker in Fig. 6-2 provides a young child with a gentle rocking action, either by the child using the handle or by an adult using a foot on the rear extension of the seat (Fig. 6-3A). The toy can be pulled like a sleigh with a rope in the forward hole (Fig. 6-3B).

The parts could all be made of 1/2-inch plywood, or you might prefer to use thicker plywood for the sides. Reinforcing strips are 3/4-inch square solid wood. All the parts are glued and nailed or screwed. The complete toy is 10 inches wide, 23 inches long, and 15 inches high.

1. With the aid of the grid of squares, set out one side (Fig. 6-3C). Cut the two sides. The bottom curve is a 24-inch radius. True the outlines and remove sharpness from all edges. Drill for the 3/4-inch dowel rod handle.
2. Attach the 3/4-inch square solid wood strips. The pair for the back should leave clearance for the seat plywood underneath.
3. Use the footboard (Fig. 6-3E) as a guide for widths of other parts. It fits against the upright strips and is shaped and drilled at the front.

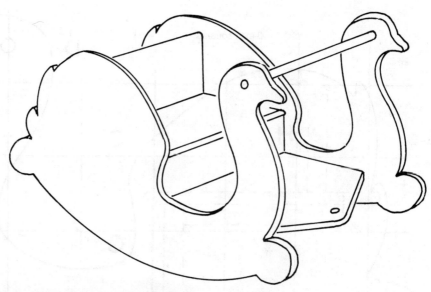

Fig. 6-2. A swan rocker can provide amusement for a toddler.

4. Make the seat (Fig. 6-3F) to the same width as the footboard. Let its front edges overhang a little and round it. Shape the rear edge to make the adult foot pedal.
5. Make the seat front (Fig. 6-3G) to fit between the other parts, and the back (Fig. 6-3H) to fit above the seat. Round its top edge.
6. Fit all these parts between the sides. During assembly, sight across to see that sides match and that the toy is going together without twist.
7. White paint would be appropriate for a swan, with black eyes at the ends of the handle.
8. To reduce the risk of unintentional sliding and to minimize wear on floor coverings, double the thickness of the bottom edge with shaped strips. Make them about 1 inch deep and glue them on. Cover the lower edges with rubber or plastic.

Materials List for Swan Rocker	
2 sides	15 × 24 × 1/2 or 3/4 plywood
1 footboard	9 × 11 × 1/2 plywood
1 seat	9 × 11 × 1/2 plywood
1 seat front	4 × 9 × 1/2 plywood
1 back	5 × 9 × 1/2 plywood
8 strips	3/4 × 3/4 × 8
1 handle	11 × 3/4 diameter

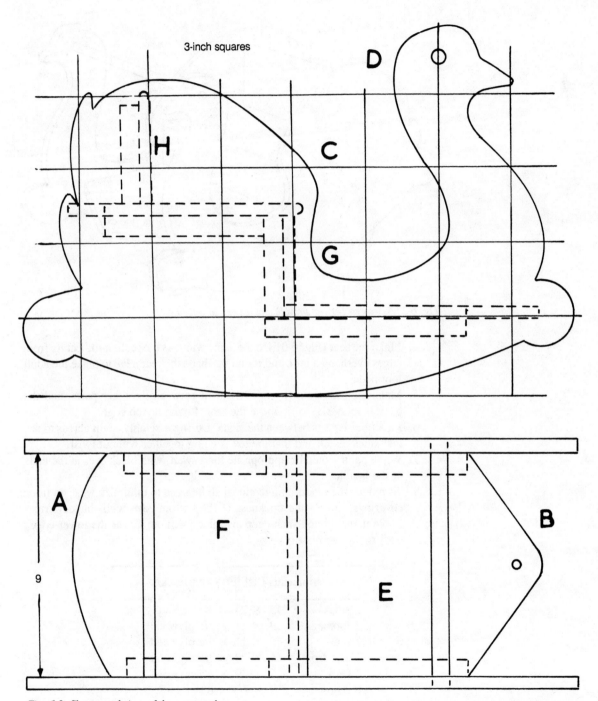

Fig. 6-3. Shapes and sizes of the swan rocker.

Ring Game

This popular game is intended to hang on a wall, so competitors can throw rings from a marked distance to try to hang them on the hooks (Fig. 6-4). It is made completely from 1/2-inch plywood.

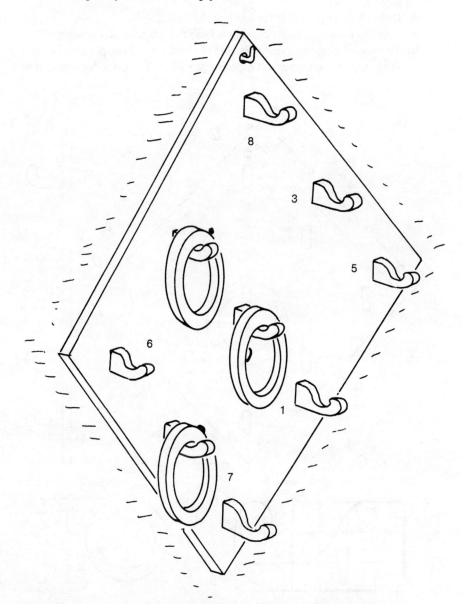

Fig. 6-4. A ring game hangs on the wall and uses plywood rings and hooks.

1. Cut the board 16 inches square and mark the centers of hooks on it (Fig. 6-5A).
2. Mark mortises diagonally at these positions 1/2 inch wide and 1 inch deep (Fig. 6-5B). Delay cutting the mortises until you make the hooks, so tenons and mortises can be fitted closely.
3. Cut nine identical hooks (Fig. 6-5C). Round the front edges and take the sharpness off all other exposed edges. Mark and cut the tenons and mortises.
4. Drill a hole (Fig. 6-5D) for hanging the board.
5. Glue the hooks in, being careful to keep them square and straight.
6. The size of ring suggested (Fig. 6-5E) should suit most players, but you might wish to make one and experiment with it. The game becomes more

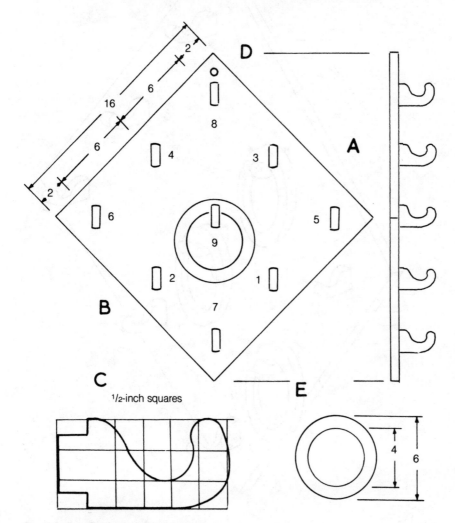

Fig. 6-5. Details of parts of the ring game.

difficult with smaller rings. The game is usually played with three rings, but you might wish to cut some spares.

7. Finish with bright paint. A light-colored board could have contrasting pegs, and the rings could all be different colors. Paint the numbers or use decals.

Materials List for Ring Game

1 board	16	× 16 × 1/2 plywood
9 hooks	1 1/2	× 4 × 1/2 plywood
3 rings	7	× 7 × 1/2 plywood

Dollhouse

A dollhouse will keep a young child occupied for hours. One problem might be its size in relation to other things in the room when it is not in use. Sometimes the dollhouse is made too small, and all its furnishings have to be very tiny and fragile if it is to scale, or the dollhouse is too big to be in proportion and looks wrong. The dollhouse in Fig. 6-6 is designed to be in reasonable proportions

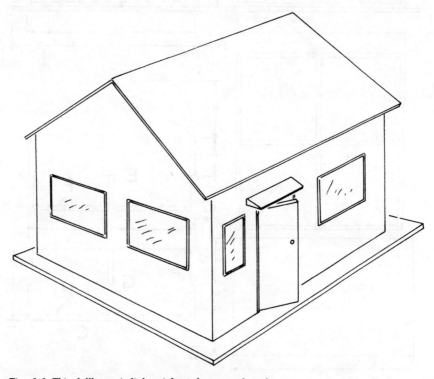

Fig. 6-6. This dollhouse is lightweight and a convenient size.

146 Toys

while avoiding too great an overall size. The house is 24 inches square on a base 28 inches square, and the height is about 18 inches (Fig. 6-7A). The scale is one-twelfth or 1-inch-to-1-foot, which is a common scale for modelmaking. This means that a bed can be 7 inches × 5 inches and a table is 3 inches high, if made true to scale. There are four rooms scaled 12 feet square, which gives scope for good furniture layouts.

The method of construction is simple, using 1/2-inch plywood, with three pieces about the same size one way (Fig. 6-7B) and three the other way (Fig. 6-7C). They notch together. Differences come in the window and door arrangements you choose, but overall sizes match. The roof is in two parts (Fig. 6-7D),

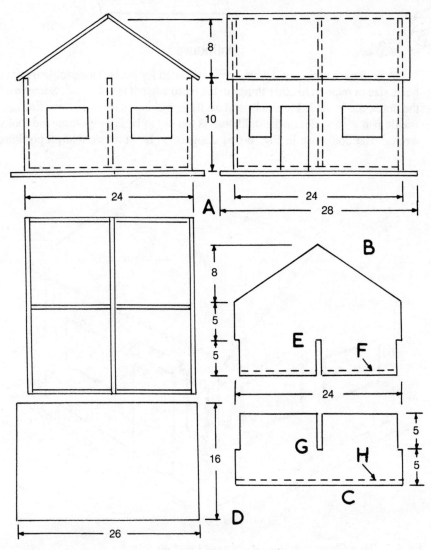

Fig. 6-7. Sizes of the dollhouse and its main parts.

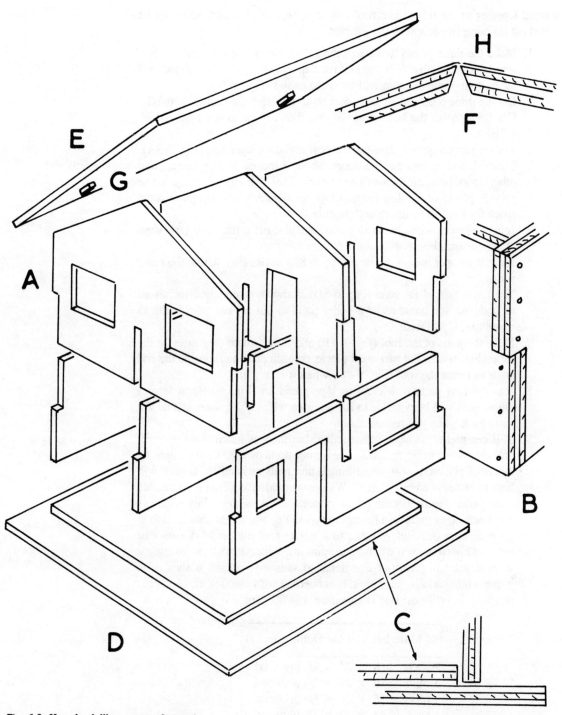

Fig. 6-8. How the dollhouse parts fit together.

hinged together at the ridge. The roof can turn back to give part access or be lifted off for complete access to the interior.

1. Make the three pieces that reach the ridge. Notch them to suit the plywood thickness (Fig. 6-7E) to half the depth of the walls. Cut $1/2$ inch off the bottom of the one that will be at the center (Fig. 6-7F).
2. Cut the three other pieces and notch them to match the first (Fig. 6-7G). Cut $1/2$ inch off the bottom of the one that will be at the center (Fig. 6-7H).
3. Try the parts together, then cut window and door openings (Fig. 6-8A). There should be two outside doors, about 7 inches $\times$ $3 1/2$ inches, and other doors between rooms. The young owner will enjoy looking in, so provide plenty of window openings in the outside walls, but leave some space for built-in furniture and pictures.
4. Join these parts with glue in all joints, and also drive fine nails both ways at the corners (Fig. 6-8B).
5. Make the upper part of the base (Fig. 6-8C) to fit easily within the outer walls.
6. The lower part of the base (Fig. 6-8D) is shown extending 2 inches all around, but you could make a wider patio or cut the extended width to very little, if you wish.
7. Make the parts of the roof (Fig. 6-8E) and miter where they meet on the ridge (Fig. 6-8F). Put two small blocks on each piece to fit inside the end walls to locate the roof (Fig. 6-8G) when it is in place.
8. Join the roof parts with cloth or tape glued on (Fig. 6-8H) to form a hinge, so it will be possible to lift one side without the other or fold the parts back when you remove them.
9. This completes the basic construction, but there is much more you will probably want to do to satisfy the young homeowner. Doors might be pieces of plywood, with small hinges and nails as handles. Details will have to be supplied by painting. Windows could be covered with flexible clear plastic, using fine strips as frames. It is possible to buy windows and many other dollhouse fittings to this scale. You might also be able to get wallpaper or adhesive paper to simulate roof tiles and brickwork or siding. Otherwise, you will have to paint in appropriate colors. You might wish to paint in plenty of fine detail to satisfy your own wishes or to impress other adults, but a child is concerned with practicalities and only needs basic facilities. Her imagination will fill gaps.

Materials List for Dollhouse

3 walls	18 $\times$ 24 $\times$ $1/2$ plywood
3 walls	10 $\times$ 24 $\times$ $1/2$ plywood
2 roofs	16 $\times$ 26 $\times$ $1/2$ plywood
1 base	23 $\times$ 23 $\times$ $1/2$ plywood
1 base	28 $\times$ 28 $\times$ $1/2$ plywood

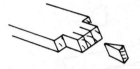

Doll Stroller

A baby stroller scaled down to suit a doll will provide plenty of enjoyment and activity for a young child. The stroller in Fig. 6-9 has a handle at a height to suit a child aged 3 or 4 years, and it will take a doll of the size she will be playing with.

Most parts are 1/2-inch plywood, and 3/4-inch square strips are used in the joints. Although the design is for a stroller for a doll, it could be altered easily to make a trolley for carrying other toys by arranging a box between the sides

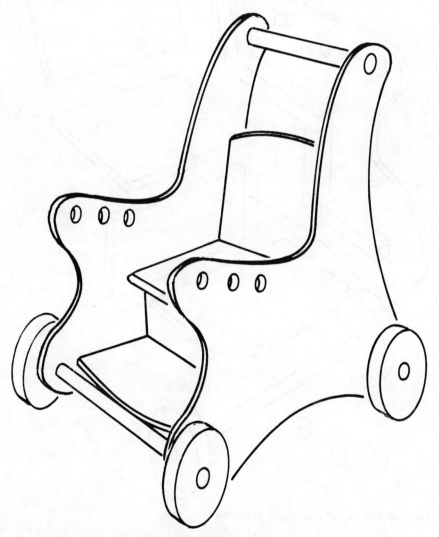

Fig. 6-9. This doll stroller is a size that a toddler can push.

150 **Toys**

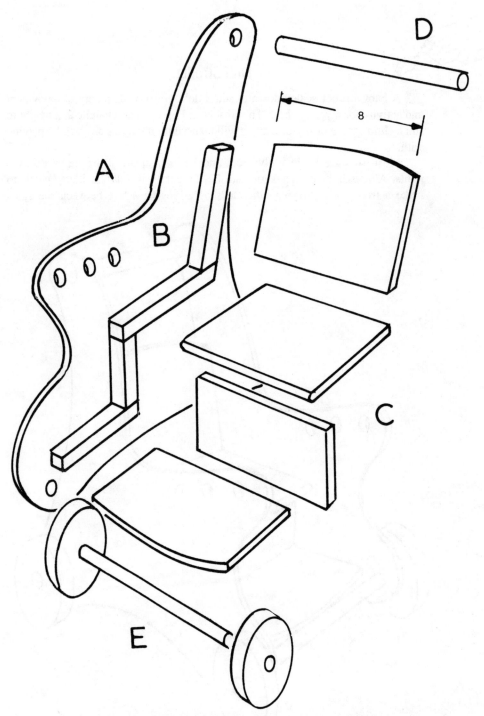

Fig. 6-10. How the doll stroller parts fit together.

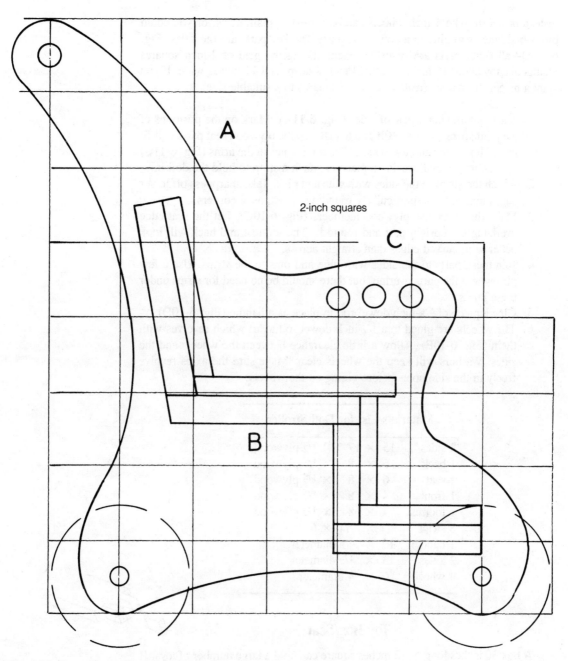

Fig. 6-11. Shape of a stroller side and its parts.

instead of a seat. The 4-inch wheels can be turned as plain discs, or you could buy wood wheels to glue to a dowel rod axle. The key parts are the sides (Fig. 6-10A)—all other parts are related to them. Using the grid of 2-inch squares results in a toy about 16 inches high, 14 inches deep, and 11 inches wide. If you want a bigger or smaller stroller, alter the squares to a suitable size.

1. Mark out and cut a pair of sides (Fig. 6-11A). Mark on the positions of the joint strips (Figs. 6-10B and 6-11B), so the plywood seat parts will fit in. Drill for the handle and axle. The three holes in the arms (Fig. 6-11C) are decorative and provide a place to attach a strap to hold the doll in.
2. Attach the strips to the sides with glue and pins. Take sharpness off lower edges and corners, particularly rounding the exposed corners.
3. Make the crosswise plywood members (Fig. 6-10C). Let the seat edge overhang the upright part and round it. The footrest and back will look better with curved edges than straight across.
4. Join these parts to the sides with glue and pins to the strips. Use a few pins where the parts overlap, but there should be no need for strips under these joints.
5. Glue a piece of $5/8$-inch dowel rod in place as a handle (Fig. 6-10D).
6. The wheels are glued to a $1/2$-inch dowel rod axle, which revolves with them (Fig. 6-10E). Allow a little clearance between the wheels and the sides. Washers will keep the wheels clear. Make sure the axles revolve freely in the side holes before gluing on the wheels.

Materials List for Doll Stroller

2 sides	15 ×	17	× 1/2 plywood
1 back	6 ×	8	× 1/2 plywood
1 seat	6 ×	8	× 1/2 plywood
1 front	4 ×	8	× 1/2 plywood
1 footrest	4 ×	8	× 1/2 plywood
8 strips	3/4 ×	3/4	× 6
1 handle	10 ×	5/8 diameter	
2 axles	13 ×	1/2 diameter	
4 wheels	3/4 ×	4 diameter	

Toy Box/Seat

A box 30 inches long × 12 inches square can hold a large number of toys. If it also serves as a seat as in Fig. 6-12, it should be popular with a child as well as an adult wanting to sit and watch him play.

The sizes suggested (Fig. 6-13A) make a seat a child can climb on that's not so low that it would be uncomfortable for an adult. The box/seat is stable, but not very heavy. Most parts are 1/2-inch plywood and 3/4-inch square stiffening strips.

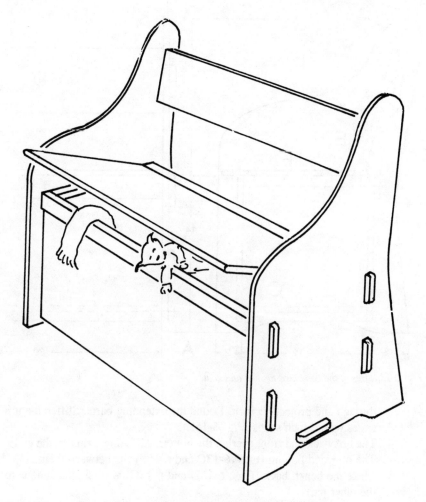

Fig. 6-12. A combined toy box and seat can be used by a child or an adult.

The length could be varied to suit your needs without altering the method of construction. Most of the lengthwise parts are tenoned into the ends and glued there. The rounded extension tenons serve as decoration. This is not intended to be a take-down project.

1. The key parts are the ends (Figs. 6-13B and 6-15A). Use the grid drawing (Fig. 6-14) to draw the outlines and mark the positions of the mortises. Cut the outlines and round the edges, but do not cut the mortises until you can also cut the matching tenons.
2. Fit stiffening strips (Fig. 6-15B and C) to come under the bottom and the lid.
3. The box back (Fig. 6-15D) and the front are the same. Cut tenons to go

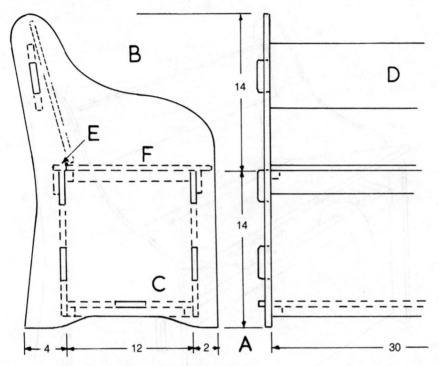

Fig. 6-13. Sizes of the combined toy box and seat.

through and project 1/2 inch. Round the extending parts. Stiffen the top edges with 2-inch strips (Fig. 6-15E).
4. The box back and front overlap the bottom stiffening strips on the ends. The plywood bottom (Figs. 6-13C and 6-15F) fits between them.
5. Make the bench back (Figs. 6-13D and 6-15G) with similar tenons to the other parts.
6. Assemble all parts made so far. Put stiffening strips under the box bottom at the back and front (Fig. 6-15H). Use glue in all joints and pins or screws where plywood comes over stiffening strips.
7. Put a strip (Figs. 6-13E and 6-15J) over the box back and its thickening piece.
8. Make the lid (Figs. 6-13F and 6-15K) to fit against this strip and project forward with a rounded edge at the front. Allow enough clearance at the ends for easy movement.
9. If possible, use a continuous piano hinge to attach the seat, otherwise put smaller hinges at about 9-inch intervals. A strip inside (Fig. 6-15L) will relieve the hinge screws of the load when anyone sits on the lid.
10. A bright paint finish will probably be best. You could add decals to the ends and front.

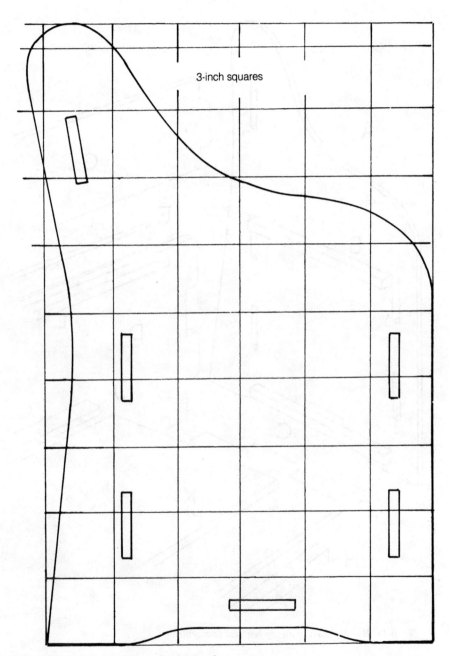

Fig. 6-14. The shape of the toy box/seat end.

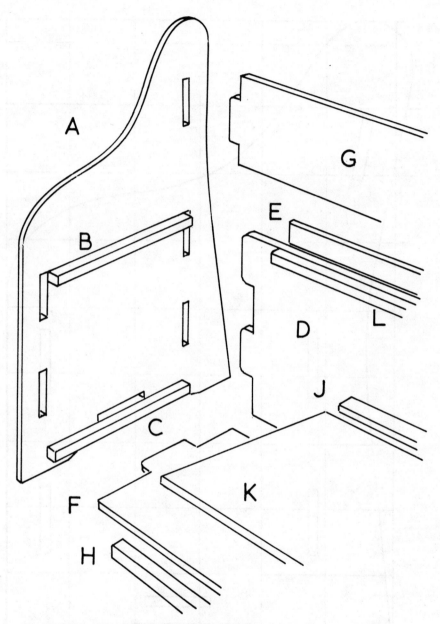

Fig. 6-15. How the parts of the toy box/seat fit together.

Materials List for Toy Box/Seat			
2 ends	18	× 28	× 1/2 plywood
1 box front	13	× 33	× 1/2 plywood
1 box back	13	× 33	× 1/2 plywood
1 bottom	12	× 33	× 1/2 plywood
2 stiffening strips	2	× 30	× 1/2 plywood
1 rear edge	1	× 30	× 1/2 plywood
1 lid	14	× 30	× 1/2 plywood
1 bench back	6	× 33	× 1/2 plywood
4 strips	3/4	× 3/4	× 13
3 strips	3/4	× 3/4	× 32

Wheelbarrow

A child likes to load toys and other things into something that allows him to wheel them about. If he sees father using a wheelbarrow in the yard, he might want one of his own. A young child will have difficulty in managing a load on one wheel, however. The wheelbarrow in Fig. 6-16A has twin wheels and is intended for a toddler. It should be light enough for him to lift and roll. He can carry a load of toys or sand and shoot it over the front. The handles are 10 inches apart and 8 inches from the floor when the wheelbarrow is standing ready for use. Check these sizes against the child who will use the wheelbarrow.

Most parts are 1/2-inch plywood. The sides are stiffened with solid wood strips. The two wheels are about a 5-inch diameter and should be obtained before you start construction in case you have to modify sizes to suit other wheels.

1. Start by making two sides (Figs. 6-16B and 6-17A). Drill holes to suit the wheel axle.
2. Make 1/2-inch-×-1-inch strips to go outside (Figs. 6-16C and 6-17B). Shape the strip ends to match the plywood, but do not drill axle holes. The strips will act as stops against the axle ends.
3. Although the wheelbarrow tapers in its length, the sides are parallel in the vertical direction. The bottom (Figs. 6-16D and 6-18A) controls the taper. Its sides are cut square, but the ends taper (Fig. 6-18B).
4. The back (Figs. 6-16E and 6-18C) is parallel and has a curved and rounded top edge. The lower edge matches the bottom.
5. The front (Figs. 6-16F and 6-18D) is also parallel, and its lower edge angle matches the bottom.
6. Join the outside strips to the sides with glue and a few pins driven from inside. Thoroughly round the handles. You might have to thin them—try the size of the child's grip.
7. If you put the bottom temporarily in position between the sides to set the angle, you can measure the length to cut the axle (Fig. 6-18E).

158   Toys

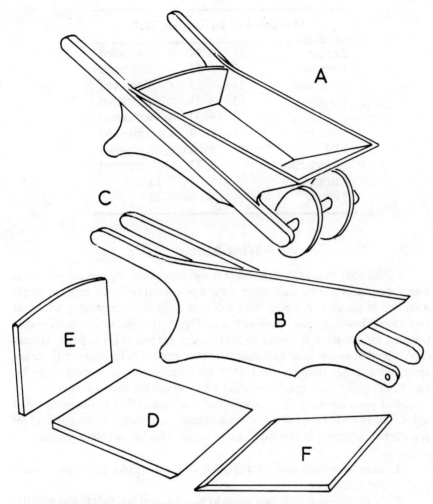

Fig. 6-16. This wheelbarrow has twin wheels to make it steady for use by a young child.

8. Make tubular spacers (Fig. 6-18F) to hold the wheels in position. The tubes need not be a close fit on the axle or tight against the wheels and could be any metal, or you could drill strips of wood.
9. Assemble the parts between the sides. Glue and nails should be satisfactory, but you could put screws near the top of each joint, where the greatest strain might be expected to come. At the same time, include the axle and wheels assembly.
10. Remove sharp edges and corners, then finish in a bright color.

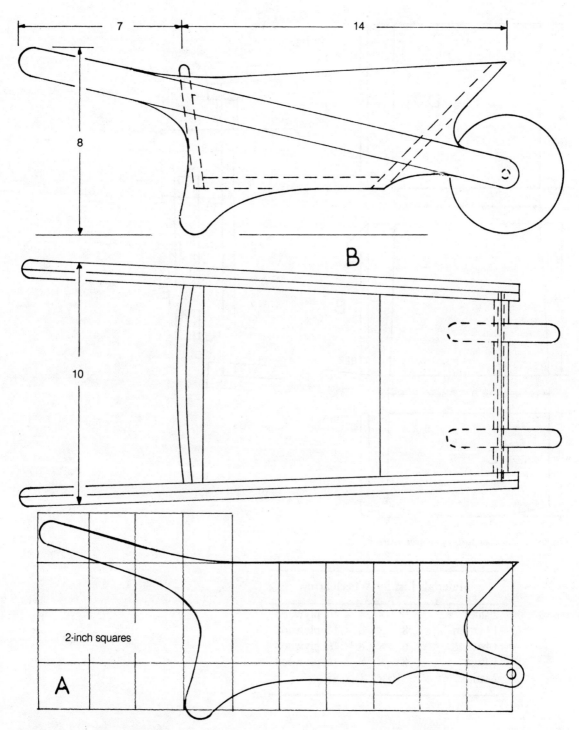

Fig. 6-17. Sizes of the wheelbarrow.

160　Toys

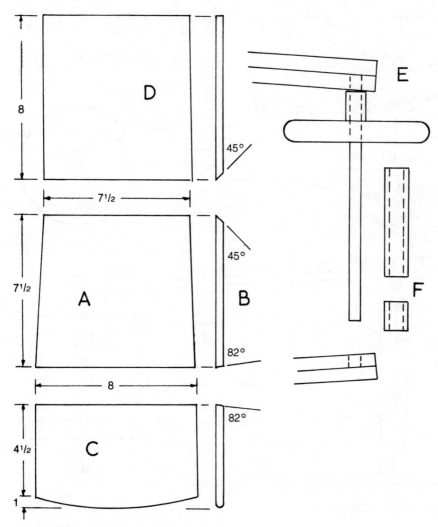

Fig. 6-18. Shapes of parts of the wheelbarrow.

Materials List for Wheelbarrow

2 sides	9 × 24 × 1/2 plywood
1 bottom	8 × 8 × 1/2 plywood
1 back	6 × 8 × 1/2 plywood
1 front	8 × 8 × 1/2 plywood
2 strips	1/2 × 1 × 25

7

Outdoor Projects

Barbecue Table/Trolley

When you set up a barbecue outdoors, you need a place to put food and equipment as well as a place for serving. Quite often you put together temporary arrangements, but they are not always satisfactory. The piece of outdoor furniture in Fig. 7-1 provides a working top or serving table that is 36 inches long, 16

Fig. 7-1. This barbecue table can be wheeled about and has a folding flap as well as shelves.

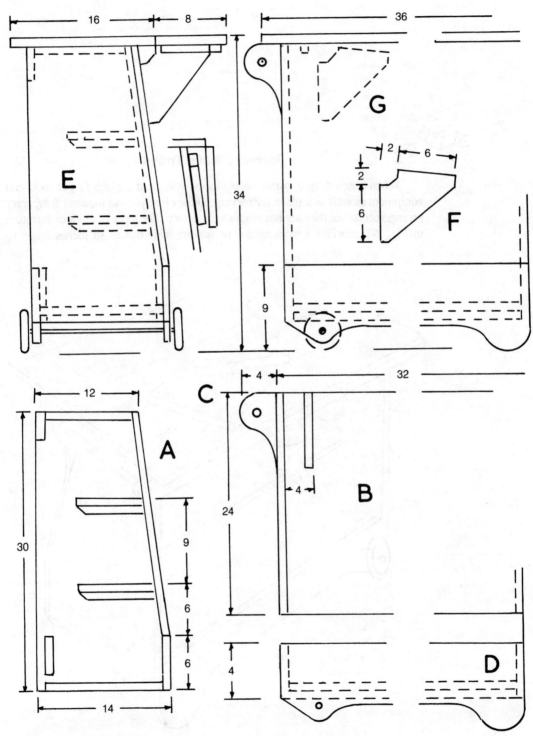

Fig. 7-2. Sizes of the barbecue table/trolley.

inches extending to 24 inches wide, and 34 inches high. Underneath, there is plenty of storage space behind an open front. There are wheels at one end and towel rails at both ends. The rail at the opposite end to the wheels also serves as a handle when you want to move the trolley.

The main parts are 3/4-inch plywood. Stiffening strips are 3/4-inch square solid wood. You could make plywood disc wheels, but store-bought, 4-inch diameter, metal-tired wheels on a 3/8-inch steel axle are appropriate for the design. Towel rails are 3/4-inch dowel rod. If you store the trolley in a dry place, you could use any plywood, but exterior grade is preferable. Use waterproof glue and nails or screws for joints.

1. Start by setting out the ends (Fig. 7-2A). The broadened bottoms give stability, particularly when pressure is put on the top flap. Attach framing strips (Fig. 7-3A). Allow for the bottom fitting under the upright strips.
2. The back (Figs. 7-2B and 7-3B) fits over the ends and extends to take the towel rails (Fig. 7-2C) at both ends. Mark on the positions of the brackets.
3. The two bottom parts (Fig. 7-3C and D) are the same, except the front piece fits between the ends and the rear piece overlaps them. The wheel size will decide depths. When the wheels are in position, the extensions at the other end (Fig. 7-2D) should hold the table horizontal. Round the top edge of the front piece. Bevel the other piece to fit against the back.
4. Make the front piece (Fig. 7-3E) to match the top edge of the back. Drill both parts to take 3/4-inch dowel rods.
5. Place stiffening strips on the bottom (Fig. 7-3F) at the back and front.
6. Assemble all parts made so far. If necessary, put a strip inside the long joint between the two parts at the back.
7. Make the shelves (Fig. 7-2E). If the plywood is not stiff enough, put strips along the front edges. You might choose to leave the shelves loose, or you can attach them to the end cleats.
8. The top overhangs 2 inches at the ends and front and 2 inches at the back, but increase this if necessary to clear the folded brackets (Fig. 7-3G).
9. Make the flap (Fig. 7-3H) to match the main top. Mark where the brackets will come and put stop strips there. Leave the top edges untreated if you will be using a painted finish, or you might prefer to add solid wood lips. You also could use a Formica or similar type top.
10. Make the bracket angle to match the slope of the back (Fig. 7-2F).
11. Hinge the two parts of the top. Four 2-inch hinges should be suitable, but keep the hinges away from where the brackets have to swing.
12. Put the top in position and screw it to the other parts. Locate the bracket positions so that they swing against the strips on the flap and close towards each other (Fig. 7-2G). Two 2-inch hinges should be satisfac-

166 Outdoor Projects

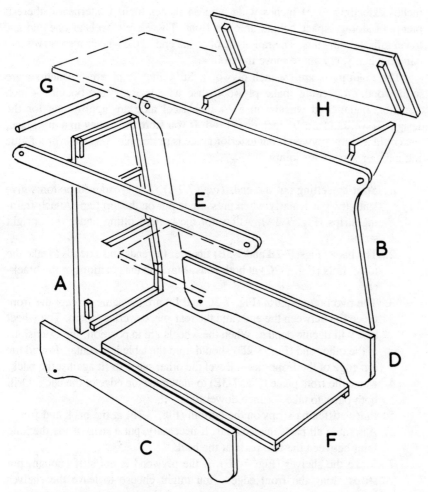

Fig. 7-3. How parts of the barbecue table/trolley fit together.

    tory on each bracket. The flap should hang closely over them when folded.

13. Glue the dowel rods in place. Fit the axle and wheels. A light-colored paint inside makes it easier to see what you have put there. A darker, durable color will be better outside.

Materials List for Barbecue Table/Trolley

2 ends	16 × 30 × ³/₄ plywood
1 back	24 × 40 × ³/₄ plywood
2 bottom parts	10 × 32 × ³/₄ plywood
1 front	5 × 40 × ³/₄ plywood
1 bottom	15 × 32 × ³/₄ plywood
2 shelves	11 × 32 × ³/₄ plywood
1 top	16 × 36 × ³/₄ plywood
1 flap	8 × 36 × ³/₄ plywood
2 brackets	9 × 9 × ³/₄ plywood
4 strips	³/₄ × ³/₄ × 32
8 strips	³/₄ × ³/₄ × 16
2 rods	16 × ³/₄ diameter

Stacking Seed Boxes

Seeds are often propagated in any container at hand, but it is better to use regular seed boxes. The design shown in Fig. 7-4A is for a box that you can make in quantity and stack as high as you like, allowing a great many to be stored safely over the floor space occupied by one box.

These boxes are 4 inches deep outside, and the gap between one and the next is 2¹/₂ inches, which is enough for most seeds to sprout to the stage where you will be moving them to the garden. Overall sizes are suggested (Fig. 7-4B), but you can make the capacity whatever you wish.

All parts are ¹/₂-inch exterior plywood. Finger corner joints are suggested (Fig. 7-4C) because they allow you to nail both ways so the corners cannot come apart. If you want to make a set of many boxes, it is advisable to cut all parts at the same time to ensure sizes match, but the instructions refer to one box.

1. Cut the sides and ends to size. Mark and cut the corner joints. Nail the parts together.
2. Cut the bottom to fit inside and nail it in place ¹/₂ inch up (Fig. 7-4D and E).
3. Make the ends to fit inside and project 3 inches (Fig. 7-4F). Round the top corners. Put ¹/₂-inch square strips (plywood or solid wood) ¹/₂ inch down. Below that drill two 1-inch finger holes at 3-inch centers (Fig. 7-4G).
4. Bevel the outer edges of the tops of the ends, if necessary, so they go easily into the bottom of the next box. Too loose a fit is inadvisable, however, if a tall stack is to stand safely.
5. You will probably leave the boxes untreated. If you do paint or use a preservative, make sure this is a type that will not affect the soil or seeds.

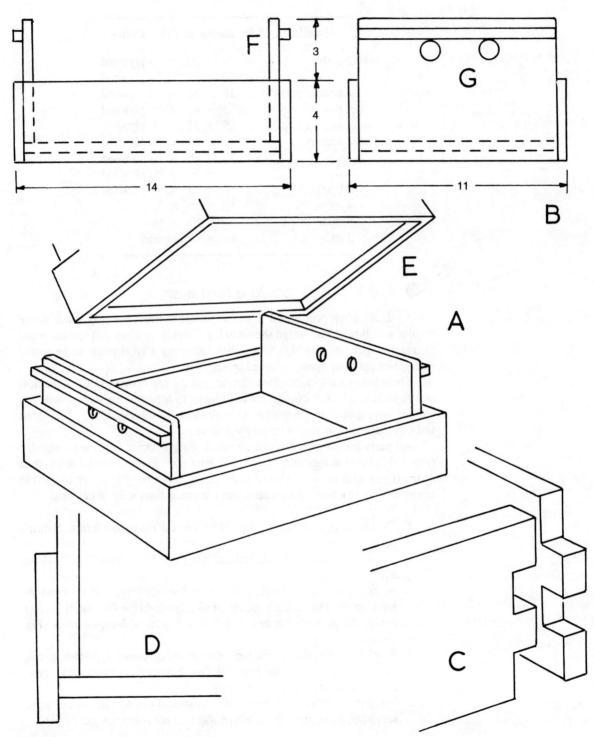

Fig. 7-4. Garden seed boxes can be arranged to stack.

Materials List for Stacking Seed Boxes

2 sides	4 × 14 × 1/2 plywood
2 ends	4 × 11 × 1/2 plywood
1 bottom	10 × 13 × 1/2 plywood
2 ends	6 × 10 × 1/2 plywood
2 strips	1/2 × 1/2 × 10

Plant Pot Stand

The ordinary garden pot is functional, but not beautiful. If you grow a plant in one and want to display it without having to transfer it to a more attractive container (which might have an unsettling effect on the plant), the pot has to be put into something that disguises its unattractive outline. The container in Fig. 7-5 is

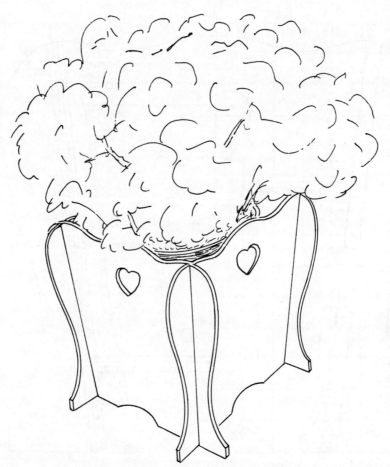

Fig. 7-5. This plywood container is intended to hold a plant in a pot.

170 Outdoor Projects

made of plywood and is drawn to fit around a pot 9 inches in diameter and height. Smaller pots would also fit, possibly with a packing under, and the design could be adapted to fit larger pots.

Most parts are 1/2-inch plywood, but for a larger container, you can increase the thickness. You could make the stand as a permanent assembly or leave the parts loosely fitted to take the stand apart and pack it flat.

1. The four sides are the same except two are slotted halfway at the top (Figs. 7-6A and 7-7A) and two are slotted at the bottom (Figs. 7-6B and 7-7B). Use the grid of 1-inch squares to mark the shape about centerlines. Cut to the outlines and sand the edges.
2. A heart cutout is shown. You might wish to omit this or make your own design, such as an initial, or do something different on each piece.
3. Try these parts together. You might have to adjust the slot sizes to fit tightly on the plywood and to come square on the stand and level at the tops and bottoms.

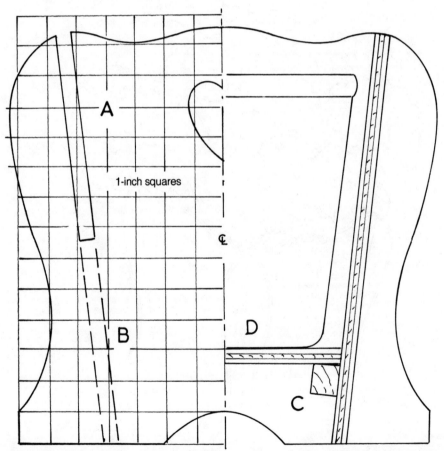

Fig. 7-6. Sizes and construction of the plant pot stand.

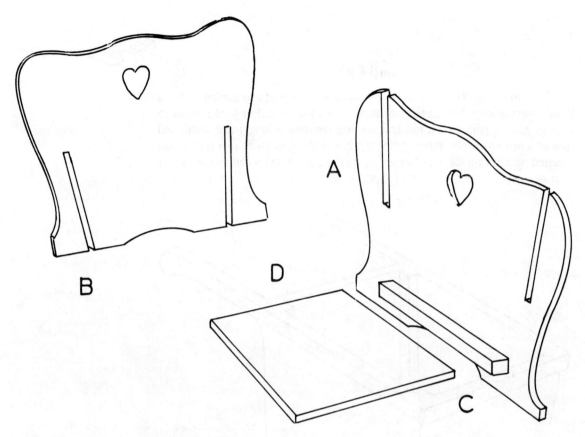

Fig. 7-7. Parts of the plant pot stand slot together.

4. The bottom support strips (Figs. 7-6C and 7-7C) need only be fitted on two opposite sides.
5. Cut a square bottom (Figs. 7-6D and 7-7D) to fit on the strips. It will hold the assembly square as well as support the pot. You could add one or more drain holes in the bottom.
6. Paint the stand in a color of your choice, to match other things on the patio or deck perhaps. You might prefer a preservative if the plant has to be watered frequently and the wood will be constantly wet.

Materials List for Plant Pot Stand	
4 sides	14 × 15 × 1/2 plywood
1 bottom	8 × 8 × 1/2 plywood
2 strips	1 × 1 × 9

Yard Cart

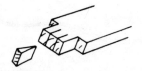

The cart in Fig. 7-8 is for moving things about the yard and garden. It has a large, deep capacity that can take anything from a load of soil or garden waste to tools or game equipment. It has handles at a convenient height and width and runs on a pair of wheels. When stationary, it is held upright with two feet. When required, you can tip the load forward or tip the cart on end when shoveling dirt in or out.

Fig. 7-8. A yard cart can carry tools and garden produce.

The main parts can be made from 1/2-inch exterior plywood. Reinforcing strips are 1-inch square. The drawings suit 6-inch diameter wheels on a 1/2-inch steel axle. You must modify the feet height if wheels of a different diameter are used. Overall size (Fig. 7-9A) is for a box 18 inches square at the top, standing 27 inches high, and with handles extending 9 inches.

1. Make two sides (Figs. 7-9B and 7-10A) using the grid of squares (Fig. 7-11) to obtain the shape. If the wheels will be a different size, alter the feet shape to bring the cart level. Mark where the bottom will come.
2. Make the bottom (Fig. 7-10B) to fit between the sides, with strips under the ends.
3. Make the back (Fig. 7-10C) and the front (Fig. 7-10D) to fit between the sides and rest on the bottom. Put strips down the sides of these parts (Fig. 7-9C) and across the bottoms.
4. Strengthen the top edges all around the outside with plywood strips 1 1/4 inches wide. Use plain pieces for the back and front (Figs. 7-9D and 7-10E). At the sides, shape the pieces to the handle outline (Figs. 7-9E

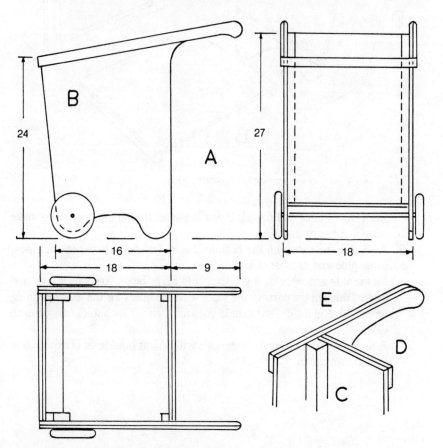

Fig. 7-9. Sizes and construction of the garden cart.

174 *Outdoor Projects*

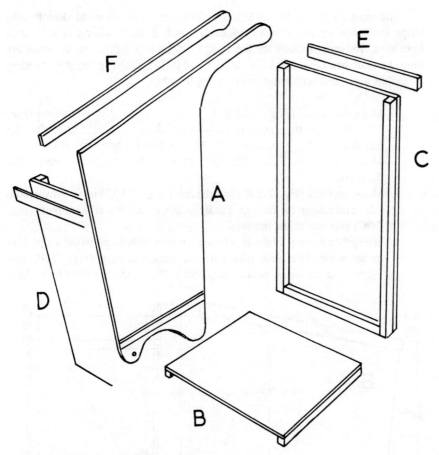

Fig. 7-10. How the parts of the yard cart fit together.

and 7-10F). Glue and nail all strips in place. Round the handles to make comfortable grips.

5. Assemble the cart with the bottom, back, and front between the sides, using glue and nails or screws.
6. Fit the axle and wheels. If you expect to carry heavy loads over long distances, thicken the parts of the sides with the holes for the axle by gluing more plywood inside, so the axle goes through a 1-inch thickness on each side.
7. Paint the wood thoroughly if the cart will be left outside or is expected to become wet often.

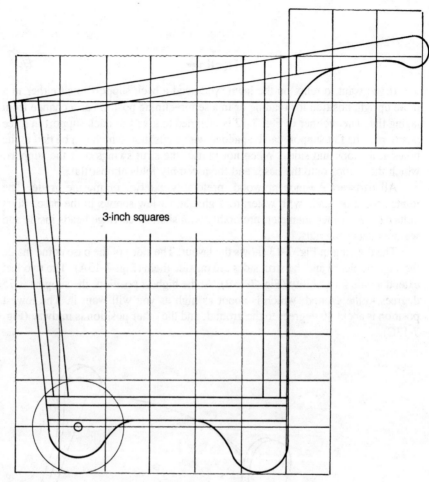

Fig. 7-11. Shape of the side of the yard cart.

Materials List for Yard Cart		
2 sides	27	× 33 × 1/2 plywood
1 back	18	× 27 × 1/2 plywood
1 front	18	× 24 × 1/2 plywood
1 bottom	16	× 18 × 1/2 plywood
2 top strips	1 1/4	× 18 × 1/2 plywood
2 handle strips	2 1/2	× 28 × 1/2 plywood
4 inside strips	1	× 1 × 27
4 inside strips	1	× 1 × 19

Recliner

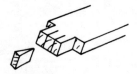

If you want to relax on the lawn, you need a back support to sit either in a more upright position for reading or in a semireclining position as a change from laying flat. The recliner in Fig. 7-12 is intended to give you back support in three positions. The flat support can be padded with a cushion, which can be tied to the holes at the tops and sides. When not in use, the strut swings over the support, which then drops onto the base, and the assembly folds almost flat.

All parts are 3/4-inch plywood, preferably exterior or marine grade. The joints should be made with waterproof glue and a few screws in the base. Pivot bolts can be 1/4-inch diameter, preferably with shallow carriage heads inside and washers under the nuts.

The drawing in Fig. 7-13 shows the layout. The sides of the base come inside the support pieces and the strut sides are outside them (Fig. 7-13A). The two feet extend a little for stability (Fig. 7-13B). At the highest position, the support is 75 degrees to the ground, which is about as high as you will want it. The lowest position is about 40 degrees to the ground, and the other position is midway (Fig. 7-13C).

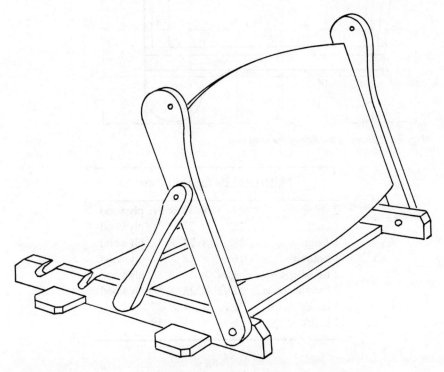

Fig. 7-12. This recliner is an adjustable backrest for use when sitting on the lawn.

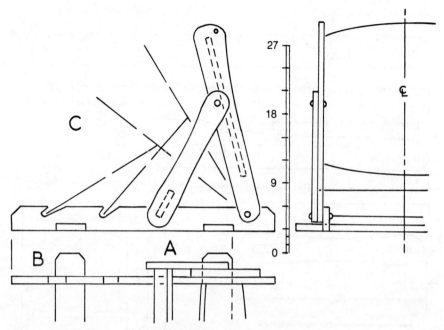

Fig. 7-13. Sizes of the recliner and the alternative positions.

1. Make two base sides (Fig. 7-14A). Position the pivot bolt holes 2 inches from the bottom edge. Estimate the angles of the notches, with their centers on the same line as the pivot hole. Allow plenty of clearance for the strut crossbar, but the notch shapes can be finalized during a trial assembly. Notch for the feet.
2. On the support sides (Fig. 7-14B), the pivot and strut holes are 1 inch from the edge and the mortises for the support are to one side of the center. Do not cut them until you can match them with the tenons on the broad support. Shape the top outline and drill the hole. Round all edges.
3. The back support (Fig. 7-14C) controls the width. Other cross member lengths should be related to it. If the side parts are not exactly 3/4 inch thick, adjust the lengths to suit. Curve the top and bottom edges of the support and round them off. Mark and cut the tenons to match the mortises in the side pieces.
4. The strut sides (Fig. 7-14D) have pivot holes and mortises for the crosspiece, which you must match to them. Hollowing the edges lightens appearance. Round all edges.
5. Check the length of the strut crosspiece (Fig. 7-14E) to bring the strut sides outside the other parts with enough clearance. Cut tenons to fit the strut sides. Round the edges.
6. The feet (Fig. 7-14F) are simple pieces with beveled corners.
7. Glue the support mortises and tenons and the strut mortises and tenons. Let the glue set, then level the outsides.

178 Outdoor Projects

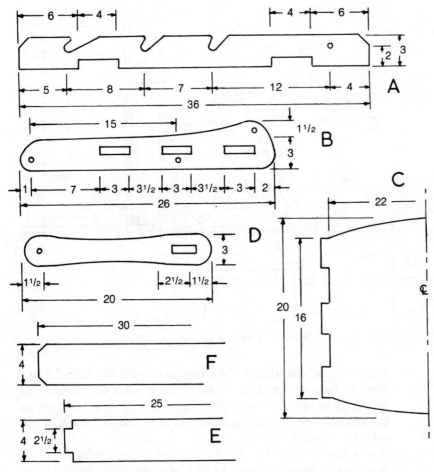

Fig. 7-14. Details of recliner parts.

8. Bolt the support sides outside the base sides and bolt the struts in place. Try the action in all positions. You might have to adjust the shapes of the strut notches in the base sides.
9. When you are satisfied with the action, put the feet in place with glue and screws. See that they hold the base sides parallel, so the other parts can fold neatly against them.
10. The recliner is unlikely to move in normal use. If you think it needs to be held down on grass, however, drill holes in the ends of each foot so that pegs or spikes can be pushed through.
11. Finish with several coats of exterior paint.

Materials List for Recliner	
2 base sides	3 × 36 × ¾ plywood
2 support sides	3 × 26 × ¾ plywood
1 support back	20 × 24 × ¾ plywood
2 strut sides	3 × 20 × ¾ plywood
1 strut crosspiece	4 × 26 × ¾ plywood
2 feet	4 × 31 × ¾ plywood

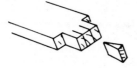

Yard Table

The table in Fig. 7-15 is for use outdoors on the lawn, deck, or patio. It is octagonal, 30 inches across, and with a socket at the center to take the upright of an umbrella or sunshade. Despite its apparent complexity, it is very easy to make.

All main parts are ¾-inch exterior or marine-grade plywood. Stiffening is provided by 1-inch square strips. If the table is well protected by paint, the plywood should be satisfactory with its edges exposed, but you could add solid wood lips to the eight edges of the top, if you wish.

The drawing in Fig. 7-16 shows suggested sizes. There are four identical leg assemblies, which overlap to leave a 6-inch square space down the center (Fig. 7-16A). The legs have flat feet (Fig. 7-16B), which spread the load on soft ground. The outside edges of the feet come under the edges of the top, so the table should stand firmly, while the arrangement of legs guards against any tendency of the assembly to wobble.

1. Set out a leg shape (Fig. 7-17A). Each shape goes 3 inches past the centerline of the table (Fig. 7-17B). A square of plywood has to fit inside the bottom (Fig. 7-17C). Put a stiffening strip down the edge (Fig. 7-17D), leaving space for the square at the bottom. This strip has to join the next leg. Mark where the adjoining leg will come (Fig. 7-17E). Put a strip across the other side (Fig. 7-17F) for joining to the tabletop. Use glue, and pins or screws in the joints. Complete four identical legs. Smooth and round all the exposed edges.

2. Make the feet 3 inches × 6 inches (Figs. 7-16B and 7-17G). Glue and screw them to the bottom of the legs.

3. Join the four legs together (Fig. 7-16C) to leave a 6-inch square space through the center.

4. Make a plywood square (Fig. 7-17C) to fit in the bottom (Fig. 7-16D). Drill for the umbrella upright—a 2-inch diameter hole should take any upright. Glue and screw this in place.

5. To obtain the shape of the top (Fig. 7-16E) start with a square and find its center. Measure the distance from a corner to the center, then step this along each edge from each corner. Join these points to draw a regular

180 Outdoor Projects

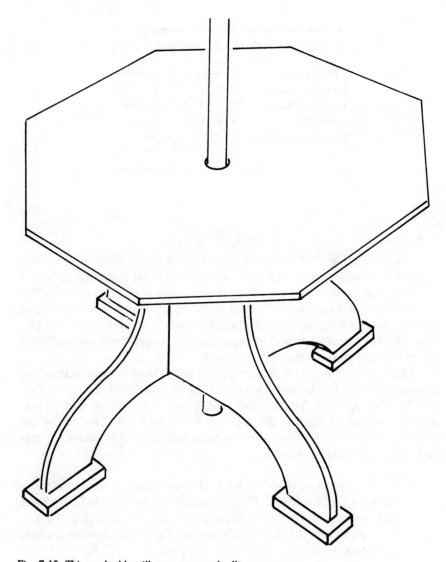

Fig. 7-15. This yard table will support an umbrella.

    octagon. Cut the shape and true the edges. Drill a hole at the center the same size as that in the bottom block.
6. Locate the top symmetrically over the legs. Glue and screw into the stiffening strips. For the best finish, sink the screwheads and cover them with plugs or stopping. Finish with adequate coats of exterior grade paint.

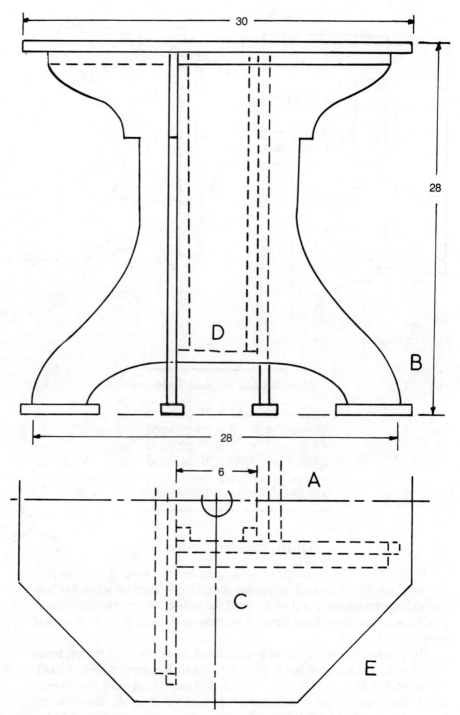

Fig. 7-16. Sizes of the yard table. Its legs are arranged around a central space.

182 Outdoor Projects

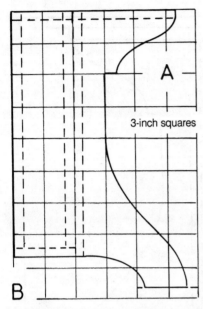
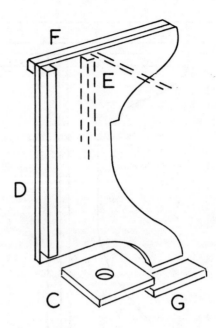

Fig. 7-17. Details of a leg of the yard table.

Materials List for Yard Table

4 legs	18 × 27 × 3/4	plywood
1 block	6 × 6 × 3/4	plywood
4 feet	3 × 6 × 3/4	plywood
1 top	30 × 30 × 3/4	plywood
4 strips	1 × 1 × 24	
4 strips	1 × 1 × 18	

Garden Kneeler/Toolbox

If you have trouble kneeling and standing again when working in the garden, or even if you do not, you will appreciate something to kneel on which has handles to help you get down and up again. If the kneeler also carries small tools, seed packets, and other small items, it becomes a valuable piece of gardening equipment (Fig. 7-18).

This project provides a kneeling space about 10 inches × 16 inches, boxes on each side, and a narrow box in front for tools such as trowels (Fig. 7-19A). You might wish to measure the space you like to kneel on and make your kneeler to suit, but be careful not to make the whole thing too large. As described, the box can be carried easily with the two handles and covers an area about 13 inches × 28 inches in use.

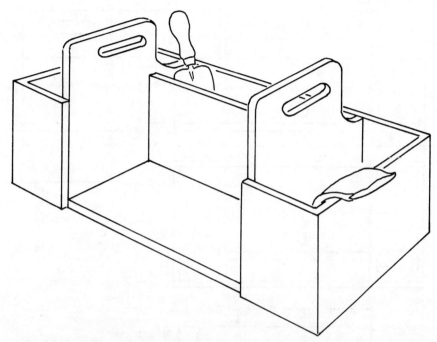

Fig. 7-18. This garden kneeler with handles has compartments for small tools and other gardening materials.

Use external-grade plywood. The handles and bottom should be 3/4 inch thick, but you can use 1/2 inch for the other parts.

1. Make two handles (Figs. 7-19B and 7-20A). Notice that the hole is offset, for balance. Round the hole edges and the top edge for comfortable grips.
2. The bottom is the same width as the handles (Fig. 7-20B). Mark out the positions of the other parts (Fig. 7-19C).
3. The outer parts fit outside the handles and the bottom. Make the front, ends, and back (Fig. 7-20C, D, and E).
4. The inner piece (Figs. 7-19D and 7-20F) goes between the handles. Round its top edge.
5. At the outer corners, you could nail the joints, preferably in a dovetail fashion (Fig. 7-19E), or you could cut finger joints and nail each way (Fig. 7-19F). Lap, glue, and nail all other joints.
6. Join the inner piece to the handles, then nail on the bottom. Nail all other parts around them.
7. Round all external edges and corners, particularly where the backs meet the kneeling space.
8. Finish with paint. You could soften the kneeler with a cushion made by covering a piece of plastic foam, which can be used loose or glued on.

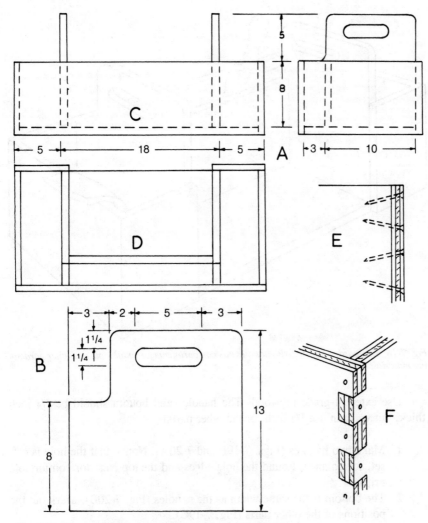

Fig. 7-19. Sizes and construction of the garden kneeler/toolbox.

Materials List for Garden Kneeler/Toolbox

2 handles	13 × 13 × 3/4	plywood
1 bottom	13 × 28 × 3/4	plywood
1 front	9 × 28 × 1/2	plywood
2 backs	6 × 9 × 1/2	plywood
2 ends	9 × 15 × 1/2	plywood
1 inner piece	8 × 18 × 1/2	plywood

Garden Kneeler/Toolbox 185

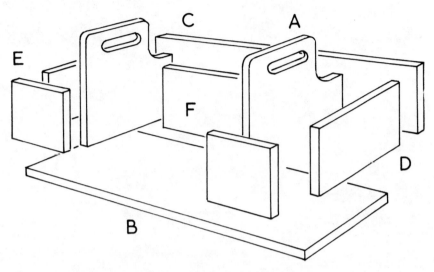

Fig. 7-20. How parts of the garden kneeler/toolbox fit together.

Index

Index

A
armchair, 59

B
barbecue table-trolley, 163
bed table, 35
bedside extending table, 45
bench seat, 70
bin, tilt, 92
blockboard, 5
bookcase, floor, 128
bookshelves, take-down, 117
box-paper roll holder, 107
boxes
    chest, 79
    garden kneeler-toolbox, 182
    paper towel holder and, 107
    stacking, 167
    tabletop, 95
    toy box-seat, 152

C
cabinets, 102
cart, yard, 172
chairs
    armchair, 59
    child-size, 55
    outdoor, 66
    recliner, 176
    side chair, 61
    swan rocker, 140
chest, 79
child's chair, 55
coffee table, 14
    take-down, 11
containers, 77-110
    box-paper towel holder, 107
    cabinets, 102
    chest, 79
    garden kneeler-toolbox, 182
    hot pads, 100
    kitchen tray, 98
    rolling tilt bin, 92
    stacking seed boxes, 167
    tabletop, 95
    tool tote box, 89
    toy box-seat, 152
    umbrella stand, 85
corner table, 41
counterbores, 8
countersinks, 8
crafts desk, 131
crib, doll's, 139

D
desk
    drafting or crafts, 131
    folding, 27
doll stroller, 149
doll's crib, 139
dollhouse, 145
dovetailed joints, 7
drafting desk, 131
dresser seat, 72

E
edge finishing, 6-7
extending table, bedside, 45
exterior grade plywood, 6

F
face veneer, plywood, 5
finger joints, 8
flap table, 24
floor bookcase, 128
fold-out top game table, 37
folding desk, 27
foot stool, 51
frames, 126

G
game table, 37

garden kneeler-toolbox, 182
glues
   joints, 7
   plywood, 5
greenhouse plant stand, 31

H

half lapped joints, 7
hardwood plywood, 4
hat and coat rack, 123
hot pads, 100

J

joints, 7-8
   dovetailed, 7
   finger, 8
   half lapped, 7
   rabbetted, 8
   stiffening, 7

K

kitchen tray, 98

M

macrame desk, 131
magazine rack, 114
marine grade plywood, 6
medium density overload (MDO), plywood, 6
mirror frames, 126
modular racks, 134
modular shelves, 134

N

nails, 7

O

outdoor chair, 66
outdoor projects, 161-185
   barbecue table-trolley, 163
   garden kneeler-toolbox, 182
   plant pot stand, 169
   recliner, 176
   stacking seed boxes, 167
   wheelbarrow, child's, 157
   yard cart, 172
   yard table with umbrella, 179

P

paper towel holder, 107
pedestal, 19
picture frames, 126
plant stand, 19
   greenhouse, 31
   potted plants, 169
plies, plywood, 4
plywood, 3-8
   blockboard, 5
   development of, 3
   edges, finishing, 6-7
   exterior grade, 6
   glues used in, 5
   hardwood vs. softwood, 4
   joints for, 7-8
   manufacturing process of, 3
   marine grade, 6
   medium density overload (MDO), 6
   plies in, 4
   quality of, 4
   sizes of, 3-4
   solid-core, 5
   strength of, 4
   surface quality, 6
   veneers, 4, 5
potted plant stand, 169

R

rabbet joints, 8
racks and shelves, 111-136
   crafts desk, 131
   drafting desk, 131
   floor bookcase, 128
   hat and coat rack, 123
   macrame desk, 131
   magazine rack, 114
   modules, 134
   picture frames, 126
   souvenir rack, 113
   take-down bookshelves, 117
   vegetable rack, 119
recliner, 176
ring toss game, 143
rocker, swan, 140
rolling tilt bin, 92
rotating top table, 24

S

screws, 7, 8
   counterbored and countersunk, 8
seats and stools, 49-75
   armchair, 59
   bench seat, 70
   child's chair, 55
   dresser seat, 72
   foot stool, 51
   outdoor chair, 66
   recliner, 176
   side chair, 61
   swan rocker, 140
   toy box-seat, 152
seed boxes, stacking, 167
shelves, 111-136
   floor bookcase, 128
   hat and coat rack with, 123
   modular, 134
   take-down bookshelves, 117
side chair, 61
softwood plywood, 4
solid-core plywood, 5
souvenir rack, 113
stacking seed boxes, 167
stand, umbrella, 85
stiffening joints, 7
stools, 49-75
   foot stool, 51
storage box, 79
stroller, doll's, 149
swan rocker, 140

T

table mats, 100
tables and stands, 9-48
   barbecue table-trolley, 163
   bed table, 35
   bedside extending table, 45
   coffee table, 14
   corner table, 41
   flap table, 24
   fold-out top game table, 37
   folding desk, 27
   game table, 37
   greenhouse plant stand, 31
   pedestal, 19
   plant pot stand, 169
   plant stand, 19
   rotating top table, 24
   take-down coffee table, 11
   yard table with umbrella, 179
tabletop containers, 95
take-down bookshelves, 117
take-down coffee table, 11
tilt bin, rolling, 92
tool tote box, 89
toolbox-garden kneeler, 182
tote box, tool, 89
toy box-seat, 152
toys, 137-160

doll stroller, 149
doll's crib, 139
dollhouse, 145
ring toss game, 143
swan rocker, 140
toy box-seat, 152
wheelbarrow, 157
trays
  bed table, 35
  kitchen, 98
  tabletop, 95
trivets, 100

trolley, barbecue table and, 163

U

umbrella stand, 85

V

vegetable rack, 119
veneers
  plywood, 4, 5

plywood, edge finishing with, 6-7

W

wheelbarrow, 157

Y

yard cart, 172
yard table with umbrella, 179